Advances in
Nutritional
Research
Volume 8

Advances in
Nutritional Research

A Continuation Order Plan is available for this series. A continuation order will bring delivery of each new volume immediately upon publication. Volumes are billed only upon actual shipment. For further information please contact the publisher.

Advances in
Nutritional
Research
Volume 8

Edited by Harold H. Draper
University of Guelph
Guelph, Ontario, Canada

Plenum Press · New York and London

The Library of Congress cataloged the first volume of this title as follows:

Advances in nutritional research. v. 1–
    New York, Plenum Press, c1977–
    1 v. ill. 24 cm.
    Key title: Advances in nutritional research, ISSN 0149-9483
    1. Nutrition–Yearbooks.
QP141.A1A3   613.2′05   78-640645

ISBN-13:978-1-4612-7891-7   e-ISBN-13:978-1-4613-0611-5
DOI:10.1007/978-1-4613-0611-5

© 1990 Plenum Press, New York
Softcover reprint of the hardcover 1st edition 1990

A Division of Plenum Publishing Corporation
233 Spring Street, New York, N.Y. 10013

Contributors

Antti Aro, Department of Epidemiology, National Public Health Institute, Mannerheimintie 166, SF-00280 Helsinki, Finland

Henry S. Bayley, Department of Nutritional Sciences, University of Guelph, Guelph, Ontario N1G 2W1, Canada

Joseph Dancis, New York University School of Medicine, 550 First Avenue, New York, NY 10016

Harold H. Draper, Department of Nutritional Sciences, University of Guelph, Guelph, Ontario N1G 2W1

S. M. Filteau, Department of Nutritional Sciences, University of Guelph, Guelph, Ontario N1G 2W1, Canada

Pirjo Pietinen, Department of Epidemiology, National Public Health Institute, Mannerheimintie 166, SF-00280 Helsinki, Finland

Paul M. Starker, Department of Surgery, Columbia-Presbyterian Medical Center, New York, NY 20032

Robin K. Whyte, Department of Pediatrics, McMaster University, Hamilton, Ontario, Canada

B. Woodward, Department of Nutritional Sciences, University of Guelph, Guelph, Ontario N1G 2W1, Canada

Preface

Volume 8 of *Advances in Nutritional Research* deals with several topics of prime current interest in nutritional research, including the role of nutrition in hypertension, in the infections associated with protein-energy malnutrition, and in pathological conditions associated with the generation of oxygen radicals in the tissues, as well as with topics of ongoing interest. Recent research indicates that reduction of obesity, of alcohol intake, and of sodium intake by salt-sensitive individuals, are the most effective non-pharmacological means of reducing high blood pressure. A new approach to therapy for infections caused by protein-energy malnutrition, based on restoration of immunocompetence by administration of thyroid and anti-glucocorticoid hormones, is presented. Current research into the role of nutrition in modulating tissue damage caused by oxygen radicals generated in various pathologies is reviewed. Two chapters deal with perinatal nutrition, one with the transfer of nutrients across the placenta and the other with the energy requirements of term and preterm infants. Another dicusses methods of assessing the nutritional status of hospitalized patients.

Contents

Chapter 1

The Transfer of Nutrients across the Perfused Human Placenta

Joseph Dancis

1. Introduction

In this discussion of *in vitro* perfusion of human placenta as an approach to the study of fetal nutrition I shall: (1) explain why, in recent years, we have discontinued using experimental animals and concentrated on human placenta; (2) describe the perfusion technique; (3) review our studies of two nutrients, glutamic acid and riboflavin, to illustrate the flexibility of experimental design that is possible and the type of information that can be obtained; (4) summarize the limitations and the potential of the perfusion technique as we have observed them.

2. Comments on Comparative Placentology

There are excellent animal models for the study of fetal nutrition. The most informative of these has been the use of indwelling catheters in the pregnant sheep. The sheep has a hypotonic uterus which permits the insertion of catheters into various maternal and fetal vessels without interrupting pregnancy. Samples of blood can be withdrawn periodically and, from their

Joseph Dancis • New York University School of Medicine, 550 First Avenue, New York, NY 10016

Advances in Nutritional Research, Vol. 8
Edited by Harold H. Draper
Plenum Press, New York, 1990

analysis, deductions can be made concerning the delivery and utilization of nutrients. As a result, a great deal has been learned about the nutrition of the fetal lamb. The problem for us is to determine what pertinence the data have for the human.

There is no doubt that there is much in common in transport mechanisms and in metabolism between the sheep and the human placenta. A fascinating feature of modern biology has been the documentation of the conservation of such fundamental attributes throughout evolution. There is, however, great variability in structure and function among placentas. A few examples will make the point.

The sheep placenta is composed of discrete cotyledonary islands. Grossly, it bears little resemblance to the compact human placenta. Its microscopic appearance is also strikingly different. Five tissue layers can be identified between the maternal and fetal circulations, contrasting with the human placenta, in which the placental trophoblast is directly exposed to maternal blood.

Given such impressive differences in morphology, it is little wonder that there are also functional differences. Glucose, the major energy source for the fetus, is transferred across the human placenta far more efficiently than in the sheep. Lactate, another significant energy source for the fetal lamb, is directed preferentially by the human placenta toward the mother. Differences in the potential for providing these nutrients suggest that there are differences in energy metabolism between the human fetus and the fetal lamb (Dancis et al., 1985b).

At the other end of the scale of molecular size, the gamma globulins of the IgG class are delivered across the human placenta at a rate that provides the newborn infant with circulating levels that are approximately equivalent to those of the mother (Hobbs and Davis, 1967). In contrast, the newborn lamb is agammaglobulinemic, receiving its gamma globulin after birth in maternal milk.

Investigators have often opted to study the rodent placenta because it is more closely related histologically to the human. Both placentas have been described as "thin" placentas, with few tissue layers interposed between maternal and fetal circulations. However, one does not have to be a skilled placentologist to detect differences in structure between the labyrinthine placenta of the rodent and the villous placenta of the human. Further compounding the difference, the rodent retains its yolk sac, which is modified to serve as a secondary organ of transfer of materials from the mother to the fetus. It is across the yolk sac placenta that IgG is transferred to the rodent fetus (Brambell et al., 1951). In the human, transfer is across the chorioallantoic placenta (Dancis et al., 1961)

These few examples emphasize the obvious. Observations made in the animal must be confirmed in the human before they can be accepted as applicable to the human. This generalization applies to all experiments in animals. It is particularly pertinent to the placenta, which is the most variable of mammalian organs.

3.   The Perfusion Technique (Schneider *et al.*, 1972)

Immediately after delivery, the placenta is examined and a suitable cotyledon is selected. The tributary fetal artery and vein are cannulated and perfusion is initiated. The corresponding decidual plate is pierced with three or four cannulas penetrating into the intervillous space and a second circulation is started. The effluent from the intervillous space pools on the decidual surface where it can be collected by suction. The finished preparation consists of an intact cotyledon or lobule with two independent circulations filling roles analogous to those of the maternal and fetal circulations.

The composition of the perfusates and other features of the perfusate system may be modified according to experimental design. The perfusate that we have used most commonly is Earle's buffered salt solution equilibrated against 95-5% O_2, N_2 and fortified with glucose and amino acids. The same basic perfusate is used for both the maternal and fetal circulations.

This experimental situation simplifies the interpretation of data in terms of placental function. However, significant deviations from *in vivo* conditions have been introduced. The maternal perfusate is injected through cannulas into the intervillous space instead of the spiral arterioles. Physically dissolved oxygen is delivered at high tensions. And the perfusates bear little resemblance to maternal or cord blood. Despite these and other differences, interesting and meaningful observations have been made.

4.   Review of Transfer Experiments

4.1   Transfer of Glutamic Acid

One class of nutrients that we initially turned our attention to was the amino acids. Studies of the uptake of several amino acids by slices of human placenta had revealed that the dicarboxylic amino acid, glutamic acid, achieved intracellular concentrations that were much higher than the others. It was not possible, however, to interpret this observation in terms of fetal welfare. It was uncertain which surface of the placental membrane -- maternal, fetal, or both -- was responsible for establishing the gradient. There was no way to know what uptake had to do with transfer across the placenta to the fetus. These questions were taken to the perfusion system (Schneider *et al.*, 1972).

The first experiment involved non-recirculated perfusates, the so-called *open:open design*. L-glutamic acid was added to either the maternal or fetal perfusate and the rate of transfer across the placenta in either direction was measured. By not recirculating the donor or recipient perfusate, a constant transplacental gradient is maintained, facilitating the calculation of transfer rates. The non-physiologic stereoisomer, D-glutamic acid, was also added to provide a measure of transfer rate attributable to simple diffusion. An L:D ratio of greater than one, signifying that the transfer rate of the L-amino acid was faster than the D-isomer, suggested a mediated transfer mechanism. We have since learned that some D-amino acids make limited use of a transport system. The interpretations that follow are not affected by this more recent observation.

The results of these studies (Table I) were unexpected and puzzling. The L:D ratio was *less* than one, contrasting with the results obtained with other amino acids that we had studied. The problem was how to explain a transfer rate for L-glutamic acid that was slower than that to be expected from simple diffusion. The explanation emerged in the next series of experiments.

The *open:closed design* was used; that is, the maternal perfusate was not recirculated and the fetal perfusate was recirculated. By not recirculating the donor perfusate, a constant level of substrate is maintained against which the fetal perfusate equilibrates. This design is particularly useful in determining the capability of the placenta to establish transplacental gradients.

Table I. Ratio of Exchange Rates of L and D Isomers[a]

	Glutamic Acid	Leucine
Maternal to fetal	0.64	1.73
Fetal to maternal	0.41	1.00

[a]L-glutamic acid is transferred across the placenta in both directions more slowly than the non-physiological D-isomer (L:D ratio <1). The L:D ratio for leucine of greater than one towards the fetus is appropriate for meeting nutritional requirements for an essential amino acid. From Schneider et al. (1979).

Table II. Transplacental gradients[a]

	Fetal/Maternal Concentrations
Alanine	1.56 ± 0.11
Glycine	1.48 ± 0.09
Lysine	1.62 ± 0.01
Aspartic Acid	0.13 ± 0.04
Glutamic Acid	0.14 ± 0.03

[a]Amino acids were added in equimolar concentrations to the maternal and fetal perfusates. The maternal circulation was kept open; the fetal perfusate was recirculated. Perfusion was for 90 minutes. Mean+SEM   N=5   From Schneider et al. (1979).

Amino acids were added in equal concentrations to the maternal and fetal perfusates. As expected, transplacental gradients were rapidly established for alanine, glycine, and lysine (Table II). In contrast, the concentrations of the two dicarboxylic acids, aspartic and glutamic acid, in the fetal perfusate fell rapidly. The only explanation in this simplified system is that placental transfer to the fetus is too slow to keep pace with uptake and metabolic consumption by the placenta.

To summarize the results of these experiments, L-glutamic acid is rapidly taken up by the placenta from both the maternal and fetal perfusates. It is, however, transferred across the placenta very slowly. As a result, in the open:open experiments the L:D ratios were less than one and in the open:closed experiments the glutamic acid concentration in the fetal perfusate rapidly fell below that in the maternal. Information such as this cannot be obtained from studies with placental slices or the more sophisticated subcellular membrane preparations. It requires an experimental preparation in which the integrity of the placenta is retained.

Interpretation of these data from the standpoint of fetal nutrition must be speculative. Glutamic acid is essential to fetal welfare as a substrate for protein synthesis and neurotransmission. Since glutamic acid cannot be supplied in adequate amounts by the mother, the fetus must find other sources. *De novo* synthesis is an obvious possibility. Another intriguing possibility as a source could be glutamine, which is readily available from the mother. The amide nitrogen of glutamic is contributed to purine synthesis, essential to the rapidly growing fetus, and the resulting glutamic acid can be used for other purposes. Excesses of glutamic acid may be returned to the placenta.

4.2   Transfer of Riboflavin

The studies of riboflavin transport will further illustrate the type of information that can be obtained with the perfusion technique. Riboflavin is a water-soluble vitamin that is essential for normal fetal development. The fetus must derive its requirement from the mother.

The initial experiments were of the open:open design (Dancis *et al.*, 1985a). ^{14}C-riboflavin and ^{3}H-L-glucose were added to either the maternal or fetal perfusates and the rates of transfer across the placenta were determined. L-glucose provided an index of the transfer rate by simple diffusion of a relatively small water-soluble molecule (molecular weight 180). It was selected for this purpose, even though smaller than riboflavin (molecular weight 376.4), because of extensive experience with it in this and other laboratories.

The results of these experiments are presented in Table III. The "transfer index" refers to the ratio of the transfer rates of riboflavin and L-glucose. A transfer index greater than one suggests mediated transfer. The transfer index towards the fetus was greater than one, but not in the reverse direction. The observations indicated a unidirectional transport system directed towards the fetus.

Table III.   Placental Transfer of Riboflavin[a]

	Maternal to Fetal	Fetal to Maternal
Riboflavin	1.65	0.65
L-Glucose	0.48	0.72
Transfer index	3.4	0.87

[a]There is evidence of indicated transfer toward the fetus (index > 1.0).   N=4   From Dancis et al. (1985a).

Table IV. Transfer Index of Riboflavin at Several Concentrations

Concentration[a]	Transfer Index
50	3.0
100	2.7
200	2.0
500	1.4

[a]ng/ml. From Dancis et al. (1985a).

Additional evidence for a transport system was sought by investigating its saturability. Still using an open:open experimental design, the concentration of riboflavin in the maternal perfusate was increased in three or four steps during an experiment. The rate of increase in transfer was proportional to the riboflavin concentration up to 100 µg/ml, above which it decreased. Transfer indices measured concurrently confirmed the impression of a reduction in the effectiveness of mediated transfer at higher concentrations, consistent with progressive saturation of a transport system (Table IV).

We next investigated whether the placenta was capable of establishing a transplacental gradient. An *open:closed* design was used. When the fetal circuit was closed, the concentration in the fetal perfusate gradually increased to levels above the maternal. When the maternal circuit was closed, the concentration in the maternal perfusate fell relatively rapidly to levels well below the fetal (Fig. 1).

In order to examine transport at the maternal and fetal surfaces of the placental membrane independently, the *bolus technique* was employed. A mixture of riboflavin and L-glucose was rapidly injected into the inflow of either the maternal or fetal circulation and frequent samples of the effluent were obtained. In six experiments, a mean of 33% of riboflavin was taken up at the maternal surface compared to 3% of L-glucose. The corresponding figures on the fetal side were 11% and 13% (Dancis et al., 1988).

In the final experiments the relative concentrations of riboflavin in the three major compartments were investigated in a closed system. Using a *closed:closed* design, equal concentrations of ¹⁴C-riboflavin were added to the maternal and fetal perfusates. After 120 min of perfusion, the radioactivity in the maternal perfusate was 846 cpm/ml, in the fetal perfusate 1586 cpm/ml, and 4104 cpm/gm in the placenta. Approximately half the radioactive riboflavin in the placenta had been converted to flavin mononucleotide and flavin adenine dinucleotide, none of which appeared in the perfusates.

The series of studies on riboflavin demonstrated how maternal riboflavin is delivered to the fetus. Riboflavin is efficiently extracted from the maternal circulation by a transport system with small capacity which is active at the maternal surface of the trophoblast. The riboflavin reaches concentrations in the placenta that are much higher than in the maternal perfusate. Part of the riboflavin is converted to nucleotides for placental use and part is released into the fetal circulation where it achieves concentrations that are higher than the maternal and lower than the intraplacental. No information was obtained on fetal requirements for riboflavin or how the capacity for placental transport relates to those requirements.

To test whether these observations had any validity *in vivo*, riboflavin concentrations were measured in paired cord and maternal sera obtained at term and a transplacental gradient was confirmed (Kirshenbaum *et al.*, 1987).

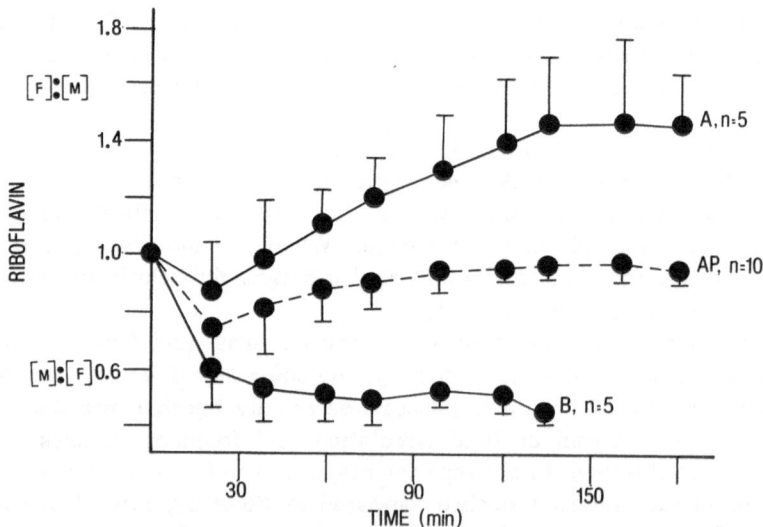

Fig. 1. Transplacental gradients of riboflavin. A = [F]/[M] ± SD with fetal circuit closed and maternal open. B = [M]/[F] ± SD with maternal circuit closed and fetal open. AP (antipyrine) = (concentration in closed perfusate/concentration in open perfusate) ± SD. Either the maternal or fetal circuit was closed (Dancis *et al.*, 1986).

5.   Potentials and Limitations

Meaningful studies of human placenta cannot be undertaken *in vivo* because of concerns for the safety of the mother and fetus. The investigator is therefore limited to tissue obtained after the termination of pregnancy. Of the many techniques that have been described for the study of this tissue, only the perfusion system maintains the integrity of the placenta.

The studies that we have presented on glutamic acid and riboflavin were selected to illustrate the experimental designs and observations that may be of interest to the nutritionist. Control of many variables is possible. By using artificially composed perfusates, particular nutrients may be added in the desired concentrations. The rate of transfer of substrates towards the fetus may be studied as well as the rate of elimination of materials from the fetal blood. The contributions of the apical and basolateral membrane to transfer can be independently studied. One may also learn about the metabolism of nutrients by the placenta and about the export of metabolites to the fetus. The role of the placenta in establishing gradients, without the confounding variables of maternal and fetal metabolism, is readily demonstrated.

Protein-binding was not a significant factor in the transfer of glutamic acid or riboflavin. Transfer rates of free fatty acids, some steroids and vitamin D are, however, greatly affected by protein-binding to albumin or to specific binding proteins. Their roles were elucidated in the perfusion system.

Our interest has been in physiological mechanisms of transport and metabolism. Others have begun to study pathological events. For example, Howard *et al.* (1987) have investigated the effect of hypoxia on the perfused normal placenta, and studies of function in the abnormal placenta have begun to appear.

Many of the advantages of the perfusion technique could be conversely presented as limitations. Excluding the mother and infant from the system makes it permissible to study the human placenta, but it also eliminates approaching the interesting problems of the interaction of the placenta with either of its partners in pregnancy. Maintaining the integrity of the placenta limits investigation into intraplacental events. We have noted the advantage of using artificial perfusates; the disadvantage is that we have introduced further deviations from the "normal."

It is clear that perfusion of human placenta has the potential for providing information that is pertinent to fetal nutrition. It is also evident that interpretation of the results requires an understanding of the limitations of the technique. The suitability of this experimental approach to any particular question must be judged by the investigator.

References

Brambell, F. W. R., Hemming, W. A., and Henderson, M., 1951, *Antibodies and Embryos*, Athlone Press, London.

Dancis, J., Lind, J., Oratz, M., Smolens, J., and Vara, P., 1961, Placental transfer of proteins in human gestation, *Am. J. Obstet. Gynecol.* **83:**167.

Dancis, J., Lehanka, J., and Levitz, M., 1985a, Transfer of riboflavin by the perfused human placenta, *Pediatr. Res.* **19:**1143

Dancis, J., Schneider, H., and Challier, J.-C., 1985b, "Nutrition of the placenta and fetus," in: *Feeding the Mother and Infant* (M. Winick, ed.), pp. 59-72, Wiley, New York.

Dancis, J., Lehanka, J., Levitz, M., and Schneider, H., 1986, Establishment of gradients of riboflavin, L-lysine, and 2-aminoisobutyric acid across the perfused human placenta, *J. Reprod. Med.* **31:**293.

Dancis, J., Lehanka, J., and Levitz, M., 1988, Placental transport of riboflavin: differential rates of uptake at the maternal and fetal surfaces of the perfused human placenta, *Am. J. Obstet. Gynecol.* **158:**204.

Hobbs, J. R., and Davis, J. A., 1967, Serum γ-G globulin levels and gestational age in premature babies, *Lancet* 1967 (1):757.

Howard, R. B., Hosokawa, T., and Maguire, M. H., 1987, Hypoxia-induced fetoplacental vasoconstriction in perfused human placental cotyledons, *Am. J. Obstet. Gynecol.* **157:**1261.

Kirshenbaum, N. W., Dancis, J., Levitz, M., Lehanka, J., and Young, B. K., 1987, Riboflavin concentration in maternal and cord blood in human pregnancy, *Am. J. Obstet. Gynecol.* **157:**748.

Schneider, H., Panigel, M., and Dancis, J., 1972, Transfer across the perfused human placenta of antipyrine, sodium and leucine, *Am. J. Obstet. Gynecol.* **114:**822.

Schneider, H., Mohlen, K.-H., Challier, J.-C., and Dancis, J., 1979, Transfer of glutamic acid across the human placenta perfused *in vitro*, *Br. J. Obstet. Gynecol.* **86:**299.

This paper is based on a presentation to the Canadian Society for Nutritional Sciences, Quebec City, June 15-18, 1988.

Chapter 2

Immunoenhancement in Wasting Protein-Energy Malnutrition: Assessment of Present Information and Proposal of a New Concept

B. Woodward and S. M. Filteau

1. Introduction

Protein-energy malnutrition (PEM) in its severe (wasting) forms is associated with impaired resistance to infectious diseases. A complex web of interrelationships is recognized among poverty, poor sanitation, malnutrition, impaired immune function and infection-related morbidity and mortality. This network of interactions is well-known in developing countries (Chandra, 1983) and may also become a major entity in industrialized nations such as the United States (Brown, 1987). In the more developed countries PEM is also becoming a problem of increasing significance among hospitalized patients as a condition secondary to chronic illnesses or trauma (Stinnett, 1983; Irving, 1985; Pinchcofsky and Kaminski, 1985). In the words of J. D. Stinnett (1983), "... medical advances present us with debilitated patients unseen a few short years ago."

B. Woodward and S. M. Filteau • Department of Nutritional Sciences, University of Guelph, Guelph, Ontario N1G 2W1, Canada.

Advances in Nutritional Research, Vol. 8
Edited by Harold H. Draper
Plenum Press, New York, 1990

Established management strategies for wasting patients with infections include nutritional support, use of antimicrobial drugs and surgical procedures such as drainage of infected sites (Sirisinha *et al.*, 1973; Meakins *et al.*, 1979; Koster *et al.*, 1981; Salimonu *et al.*, 1983). At present, however, severely malnourished subjects commonly succumb to infections despite vigorous medical intervention. Antibiotics, for example, may fail to control infections in malnourished subjects with high pathogen load (Murray and Murray, 1977). This could be, in part, because these drugs often function *in vivo* primarily as bacteriostatic agents (Boyd and Sheldon, 1984) and because they depress some immunological defence mechanisms even further (Tarnawski and Batko, 1973; Munster *et al.*, 1977; Sheng *et al.*, 1987; Beris and Miescher, 1988). Moreover, it is often impossible to reverse the wasting process in malnourished subjects until associated problems of sepsis (Irving, 1985) or severe trauma (Frayn, 1986; Gusberg, 1986) have been dealt with. From these considerations, therefore, it appears important to explore and develop new possibilities for improving disease resistance in severe PEM without the necessity of first achieving a reversal of the wasting process. Friedman (1987) has recently indicated the importance of this goal with reference to severely wasted men. Olusi *et al.* (1980) and Torun *et al.* (1981) also have discussed the importance of accelerating the recovery of immune functions in severely malnourished patients.

The purposes of the present article are, first, to analyze available evidence with regard to immunoenhancement as a rational therapeutic measure during continued severe PEM and, second, to highlight a new concept on which to base future research relating to this subject. Others have alluded to the novelty and potential of immunoenhancement therapies for malnourished subjects (Chandra and Newberne, 1977; Olusi *et al.*, 1980; Torun *et al.*, 1981; Moldawer *et al.*, 1984). The present authors, however, are unaware of any long-term commitment to the investigation of such therapies or to the development of their conceptual foundation. Consequently, the information reviewed herein is largely preliminary in nature.

2. Supplementation With Single Nutrients

PEM is a systemic disease which involves apparent deficiencies of numerous vitamins and trace minerals (Dowd and Heatley, 1984). The importance of micronutrients in immune functions is widely recognized (Gershwin *et al.*, 1985). Many investigators, therefore, have suggested that immunological impairments in PEM may be attributable to micronutrient deficiencies (Hansen *et al.*, 1982; Richter, 1982). A corollary to this suggestion is that supplementation with one or a few micronutrients will, apart from other therapeutic efforts, improve immune functions in malnourished subjects. Most research on this idea has been directed towards zinc supplementation.

2.1   Zinc

Low serum or plasma zinc levels frequently occur in severe PEM (Halstead and Smith, 1970; Khalil et al., 1974; Golden and Golden, 1981a; Filteau and Woodward, 1982, 1984), and low zinc concentrations have also been reported in human liver (Warren et al., 1969) and erythrocytes (Khalil et al., 1974) in this condition. Plasma zinc levels within the normal range, however, can also occur in severe PEM (Golden and Golden, 1981a; Castillo-Duran et al., 1987). Such seemingly contradictory findings may reflect differences in the type of nutritional deprivation, i.e., marasmic or edematous (Golden and Golden, 1981a), variable exposure to infectious and other traumatic agents (Solomons, 1979) and possibly variable zinc intakes (Hansen et al., 1982). The true significance of plasma zinc concentrations in PEM is therefore unclear. This is important to recognize while considering evidence concerning the efficacy of zinc as an immunostimulant in PEM.

Castillo-Duran et al. (1987) gave oral zinc supplements to marasmic children (2 mg Zn/kg, as zinc acetate, daily) throughout 90 days of rehabilitation. Supplemental zinc (over and above an intake of 3-3.5 mg/day supplied by the rehabilitation formula) reduced the number of patients unable to generate a cutaneous delayed hypersensitivity response either to the tuberculosis antigen (PPD) or to Candida, and reduced the incidence of "clinically significant" skin infections. In an earlier investigation, oral zinc supplements (2 mg Zn/kg body weight, as zinc acetate, daily for 10 days) appeared to stimulate thymus re-growth, determined by chest radiographs taken before and after the supplementation period, when given to children following rehabilitation from PEM (Golden et al., 1977). The subjects had received a high-energy liquid formulation supplying 0.5 mg Zn/kg body weight for at least four weeks prior to the study, and had achieved expected weight-for-height. Nevertheless plasma zinc levels were low and thymic atrophy remained at the time that zinc treatment was initiated. The results were interpreted to suggest that the thymolymphatic atrophy and cell-mediated immunodepression of PEM are caused primarily by zinc deficiency. Both these results and the findings of Castillo-Duran et al. (1987), however, could reflect simply the high zinc requirement demonstrated by Golden and Golden (1981b) and more recently by Morgan et al. (1988) and by Simmer et al. (1988) in association with the rapid growth phase of recovery from prepubescent PEM.

In a preliminary study with malnourished children (Golden et al., 1978), topical administration of 1% zinc sulphate increased the cutaneous delayed hypersensitivity response to Candida at the sites of zinc application.

In each subject the magnitude of response at the site of zinc application was compared to that achieved at a distant site which did not receive exogenous zinc. The investigation was initiated within two days of admitting the subjects into hospital so that a maintenance therapeutic regimen, presumably including a liquid diet similar to that of Golden and Golden (1981a), was provided before and during the two days of the study. This feeding protocol provides 0.4 MJ gross energy, 0.6 g protein and 0.1 mg Zn per kg body weight daily. The results were considered to suggest that an adequate supply of dietary zinc would, apart from other nutrients or energy, promote improved immune functions in PEM. The response obtained from topical zinc, however, may reflect nothing more than a pharmacological action requiring zinc concentrations unattainable systemically. For example, supra-physiological concentrations of zinc are mitogenic for thymic lymphocytes and thymocytes (Nordlind, 1985; Warner and Lawrence, 1986) and, under suitable conditions (Malavé and Benain, 1984), can also enhance the proliferative response of thymic lymphocytes to polyclonal mitogens. The mitogenic influence of zinc also appears quite nonspecific insofar as the same effect can be produced by similar concentrations of other divalent cations (Warner and Lawrence, 1986). Nevertheless, it must be acknowledged that Golden *et al.* (1978) achieved zinc-induced immunostimulation in the face of continued energy deprivation and major deficits of many nutrients in addition to zinc.

The efficacy of zinc as an immunostimulator in PEM, when administered without prior or concurrent nutritional therapy, has been examined in only one study (Filteau and Woodward, 1984). In this work, zinc was given parenterally to weanling mice (2.2 mg Zn/animal, subcutaneously in oil, weekly) throughout the period of malnutrition. The treatment maintained normal serum zinc levels in a model which otherwise caused a dramatic depression in this parameter, but did not influence the primary antibody response to sheep red blood cells or the delayed hypersensitivity reaction to this antigen. Both immune responses remained depressed, and splenic and thymic indices also remained unaffected and low. It is tempting to conclude that zinc supplements may be of little benefit in malnourished subjects whose clinical problems preclude rapid improvements in nutritional status. Two cautionary points, however, must be made concerning this work. First, Filteau and Woodward (1984) measured the end-product of complex, multi-component immune reactions. Such an experimental approach could fail to reveal zinc-related improvements in isolated components of the total response. Second, the sole measure used to quantify the antibody response was serum titre at a single time point following immunization. This approach is insensitive to differences in the kinetics of the antibody response and in the turnover rate of secreted antibody.

Finally, it should be noted that zinc appears to be an immuno-modulator for which a unique optimal level of supplementation exists in the case of each immune function. Moreover, immunodepression can occur at high but realistic levels of zinc administration. Evidence for these points derives from both *in vivo* and *in vitro* studies, and is briefly summarized by Filteau and Woodward (1986). With regard to infectious disease resistance, an optimal level of zinc supplementation is also apparent, and benefits may depend on the type of infecting organism and the timing of zinc treatment (Mansour *et al.*, 1983; Tocco-Bradley and Kluger, 1984). Any therapeutic application of zinc supplements, therefore, should be made with caution and in an individualized manner.

2.2   Nutrients Other Than Zinc

A copper supplement (80 μg Cu/kg, as copper sulphate, daily) reduced the total number of "severe" lower respiratory infections when given to marasmic infants throughout a 90-day rehabilitation period (Castillo-Duran *et al.*, 1983). Other infectious disease-related parameters were not affected and measures of immune function were not made. It remains unclear, there-fore, whether an immunological mechanism was involved and whether copper supplements can be effective in the face of continued wasting disease.

In a study of adolescent rats fed a low-protein diet which limited growth rates and induced low serum albumin levels, Ahmed and Qadri (1985) reported lower serum antitetanus antibody levels in response to tetanus toxoid than were found in animals fed the same diet supplemented with lysine, lysine plus methionine, or lysine plus tryptophan. The total protein level of the deficient diet (7%) permitted a growth response to supplementation with lysine, the first limiting amino acid (by calculation based on the formulation presented), but not to supplementation with methionine (second limiting) or tryptophan (third limiting) in addition to lysine. Interestingly, however, either methionine or tryptophan given together with lysine appeared to increase the anti-tetanus antibody response (serum titre) over that achieved with a lysine supplement alone. This finding is of interest in view of the well-known importance of methionine in initiation of eukaryotic protein synthesis (Ottaway and Apps, 1984) and in view of studies (Sidransky, 1986) suggesting an important role for tryptophan in promoting hepatic mRNA synthesis and translocation. It is presently unknown whether similar results could be obtained in wasting malnutrition or by administering methionine or tryptophan following establishment of severe immunodepression.

No additional information appears to exist as to the influence of single amino acid supplements on immune functions or disease resistance in PEM.

Immunostimulant activities of orally administered lysine and arginine supplements, however, are reported in well-nourished young adult and senescent rodents (Barbul *et al.*, 1977, 1980, 1983; Fabris *et al.*, 1986). Arginine, at least, appears to function by way of the hypothalamic-pituitary axis (Barbul *et al.*, 1983), a physiological system which is generally intact even in severe PEM (Becker, 1983). The potential of this amino acid as an immunostimulant in PEM therefore may be worthy of investigation.

3.   Adoptive Immunotherapy

Each immunological function, not surprisingly, may exhibit unique sensitivity (or resistance) to impairment by severe PEM (e.g., Gross and Newberne, 1980). Moreover, there is excessive potential (redundancy) in many immune functions so that immunological impairments often must be large in order to be physiologically meaningful. Taken together, these concepts suggest that replenishment with one or a few immunological components which are particularly severely diminished might improve disease resistance substantially in PEM.

3.1   Transfer Factor

Transfer factor is a crude, dialyzable extract of leukocytes which contains a variety of components capable of nonspecific augmentation or suppression of immune functions but which also can be used for passive transfer of long-lasting, specific, cell-mediated immunity (Fudenberg, 1985). Jose *et al.* (1976) reduced the number of episodes of diarrheal disease by giving two subcutaneous injections of transfer factor, three months apart, to patients in an advanced stage of rehabilitation from PEM. Prophylactic value was apparent, therefore, from this strategy of treatment. No protection was achieved against chest, middle-ear or skin infections, but the anticipated immunological specificities of the transfer factor preparation were not clearly defined.

Subcutaneous injection of a crude saline extract of white blood cells taken from a strongly tuberculin-positive, healthy adult converted tuberculin-negative marasmus and kwashiorkor victims to tuberculin-positive within 48 hours (Brown and Katz, 1967). This result is particularly interesting because it was achieved immediately upon admission of the malnourished subjects to hospital before reversal of wasting was apparent. Effectiveness with regard to disease resistance, however, was not evaluated nor was the immunological specificity of the transferred immune reactivity demonstrated. Furthermore, carrier-injected controls were not included in the study, presumably because an experimental investigation was not the primary purpose of the work. Nevertheless, other studies (cf. Walker *et al.*, 1975) render unlikely the possibility that such a rapid and early improvement in tuberculin response is attributable to the initial nutritional therapeutic measures employed. Walker *et al.* (1975) also treated marasmus and kwashiorkor victims with intradermal injections of transfer factor at an early stage of rehabilitation. This careful study involved use of carrier-injected controls and demonstrated that the preparation used could transfer tuberculin reactivity to a previously unreactive, healthy recipient. The preparation failed to transfer to malnourished subjects either specific immunity to tuberculin or nonspecific augmentation of immune responsiveness to two additional recall antigens.

The divergent results of Walker *et al.* (1975) and Brown and Katz (1967) may reflect the different protocols followed for preparing injected extracts. This possibility serves to emphasize a major disadvantage to the experimental or therapeutic use of transfer factor, viz., the ill-defined nature of its active component(s). No adverse effects of transfer factor injections were reported in the cited studies. The efficacy of this treatment in PEM remains unproven, but interest has shifted towards other immunologically-active preparations, many of which are chemically defined entities.

3.2 Thymic Hormones and Extracts

Thymic epithelial cells produce soluble factors, including peptide hormones, which appear to promote intrathymic T lymphocyte differentiation (Auger *et al.*, 1982; Schuurman *et al.*, 1985) and which may also help to maintain mature extrathymic immune functions (Zatz and Goldstein, 1985). In addition, direct contact between thymocytes and the thymic epithelium appears important in T cell differentiation (Itoh *et al.*, 1982). In severe PEM, involution of the thymic epithelium occurs (Mittal *et al.*, 1988), and pathology is evident in the hormone-secreting vacuolar apparatus in the remaining epithelial cells (Mittal and Woodward, 1985, 1986). Moreover, serum bioactivity of the thymic hormone thymulin is generally low in severe PEM (Chandra, 1979; Mittal *et al.*, 1988). Therapeutic use of thymic hormones, therefore, might be expected to enhance the rate at which the depleted immune system is replenished in malnourished patients.

Jackson and Zaman (1980) found a low percentage of mature T cells, as assessed by erythrocyte rosetting analysis, in the peripheral blood of malnourished children. Peripheral blood mononuclear cells from these patients, however, exhibited a near-normal percentage of T cells when cultured with a supranormal level of the thymic hormone thymopoietin (cf. Goldstein and Audhya, 1985). The percentage of rosette-positive cells was not influenced in cultures from well-nourished subjects. Similar results were reported by Olusi *et al.* (1980) and by Cruz *et al.* (1982) except that these investigators used a crude thymic extract (thymosin fraction 5). Interestingly, the latter material contains only traces of thymopoietin (Goldstein *et al.*, 1981), the effector compound used in the investigation of Jackson and Zaman (1980). In the presence of appropriate stimuli, and in a favorable milieu (studies were not done with media containing autologous serum), T lymphocytes from malnourished subjects therefore appear able to undergo differentiation to maturity.

Investigations are needed in which thymic hormones are administered *in vivo* with assessment of lymphocyte-mediated functions. Such studies have been initiated in the laboratory of Watson using a mouse model of wasting protein deficiency. In this experimental system, injections of thymosin fraction 5 resulted in improved splenocyte responses to the B cell mitogen *E. coli* lipopolysaccharide, and to the T cell mitogen phytohemagglutinin (Watson *et al.*, 1983; Watson and Lim, 1986). Importantly, thymosin injections were begun after depressed mitogen responses had developed in the malnourished animals, so that immunorestoration was demonstrated. An interesting clue to the mechanism of the phenomenon was provided in an enhanced mitogenic response correlated with a modest reduction in serum corticosterone concentrations (which, nevertheless, remained above normal).

Watson's group also has investigated the influence of injected thymosin fraction 5 on a number of cell-mediated immune functions assessed either *in vivo* or *in vitro* in models of moderate protein deficiency which appeared to permit positive nitrogen and energy balance (Petro *et al.*, 1982; Petro and Watson, 1982). The regimen of moderate protein deficiency, itself, increased some of the immune functions evaluated, and thymosin generally effected, at best, a modest further immunostimulation. Significantly, however, T cell-mediated resistance to the pathogen *Listeria monocytogenes* was impaired by the protocol of moderate protein deficiency and was restored by thymosin, at least in aging mice fed the low protein diet (Petro *et al.*, 1982).

In the same series of experiments, both well-nourished and "mildly deficient" mice responded inconsistently to exogenous thymosin and sometimes exhibited immunodepression (Petro and Watson, 1982; Watson *et al.*, 1983; Watson and Lim, 1986). It will be important, therefore, to determine whether any potential benefits from therapeutic use of thymic hormones or extracts are related inversely to the degree of wasting. The active thymus-derived molecules must also be identified, inasmuch as the use of ill-defined extracts is conceptually undesirable. Moreover, the various active components may be effective only in the correct combination (Anon., 1983). Assay technology for thymic hormones is improving rapidly and should provide an important foundation for mechanistic investigations into the therapeutic value of these compounds in both primary and secondary immunodeficiency conditions.

3.3   Interleukin-1

Interleukin-1 (IL-1) is a peptide hormone which is synthesized primarily, but not exclusively, by mononuclear phagocytes and which has numerous, diverse functions important in integrating innate and specific immunity (Dinarello and Mier, 1986). Synthesis and/or secretion of IL-1 bioactivity is low in cultures of mononuclear phagocytes from malnourished human subjects (Hoffman-Goetz et al., 1981; Bhaskaram and Sivakumar, 1986; Kauffman et al., 1986) and experimental animals (Bell and Hoffman-Goetz, 1983; Bell et al., 1985; Moldawer et al., 1985). Moreover, beneficial IL-1-dependent responses to infection are attenuated in malnutrition, e.g., fever hypoferremia, hypozincemia, and changes in serum levels of acute phase proteins (Hoffman-Goetz and Kluger, 1979a, 1979b; Bell and Hoffman-Goetz, 1983; Moldawer et al., 1985; Kauffman et al., 1986; Bradley et al, 1987). The theoretical basis for expecting benefits from administering exogenous IL-1 to malnourished patients has been outlined recently in some detail (Moldawer et al., 1984) and some experimental support is now available for this idea.

Purified human IL-1 exhibited immunorestorative activity in severely protein-deficient guinea pigs insofar as this compound improved blood clearance of an intravenously administered dose of living *Pseudomonas aeruginosa* (Moldawer et al., 1985). The effect was not attributable to enhanced blood clearance, and was hypothesized to result from increased neutrophil microbicidal activity. In the same study, the malnourished guinea pigs also were able to respond to exogenous IL-1 by developing hypoferremia and hypozincemia, a result confirmed in a later investigation (Drabik et al., 1987). In other work, wasting, hypoalbuminemic rabbits fed a low-protein diet retained the ability to exhibit a fever response to exogenous IL-1-containing supernates (Hoffman-Goetz and Kluger, 1979a, 1979b). In addition, peripheral blood T cells from wasting, protein deficient rabbits exhibited an increase in concanavalin A-induced mitogenesis when cultured with IL-1-containing monocytic supernates (Bell et al., 1985; Hoffman-Goetz et al., 1985), although it is not yet clear whether IL-1-dependent T cell functions could exhibit similar responsiveness *in vivo* in otherwise untreated PEM.

In several studies, however, IL-1-dependent antimicrobial responses (fever, neutrophil leukocytosis and the acute-phase protein response) have remained absent or attenuated in severely malnourished animals given supplements of IL-1 or IL-1-containing extracts (Bell and Hoffman-Goetz, 1983; Moldawer et al., 1985; Drabik et al., 1987). Under some conditions, therefore, a degree of metabolic recovery may be necessary to permit a response to exogenous IL-1 in wasting PEM. Moreover, this is not simply a matter of enhanced availability of energy and amino acid nitrogen (Drabik *et al.*, 1987). Mechanistic investigations into this problem should suggest means of improving attenuated responses to IL-1 in patients whose illnesses preclude rapid reversal of the wasting process. IL-1, therefore, holds promise as an immunostimulant in PEM, and this is particularly so now that recombinant human IL-1 is available (cf. Dinarello and Mier, 1986).

Evidence regarding the immunostimulant activity of cytokines other than IL-1 in intact malnourished animals appears ⁺o be unavailable. Preliminary *in vitro* studies (Petro and Wess, 1987), however, suggest that interleukin-2 (IL-2) may be worthy of investigation. By contrast, interferon failed to yield encouraging initial results (Salimonu *et al.*, 1983). In the future it will be of interest and importance to determine the effectiveness of the colony-stimulating factors. These hormones induce production of differentiated leukocytes by the bone marrow and have yielded spectacular early successes, without obvious debilitating side-effects, when used therapeutically in victims of some types of cancer, AIDS and end-stage kidney disease (Sachs, 1987). The genes for the human colony-stimulating factors have been cloned so that chemically-defined therapeutic material now can be produced.

4.   Manipulation of Classical Endocrine Hormones: A New Concept

Aschkenasy (1957) originally proposed that the immune impairments which occur in PEM result from "hormonal disturbances" associated with this condition. Since that time the neuroendocrine regulation of immune functions has become widely recognized (e.g., Besedovsky *et al.*, 1985), and much information has become available which documents the endocrine changes in PEM (Becker, 1983). Although these endocrine responses may best be viewed as adaptive (Becker, 1983), they may also exact a cost which includes immunodepression. It follows, therefore, that one possible short-term mode of immunotherapy in PEM is manipulation of blood levels of immunologically active hormones and such a strategy could prove effective apart from any changes in nutrient or energy intake. This concept has only recently been delineated clearly (Filteau *et al.*, 1987a, 1987b; Filteau and Woodward, 1987b; Perry *et al.*, 1988). It currently derives some experimental support from encouraging initial results with triiodothyronine (T_3) and an anti-glucocorticoid strategy.

4.1   Triiodothyronine

Thyroid hormones regulate numerous immune functions (Fabris, 1981; Filteau, 1987). Blood levels of total and free T_3 are consistently low in severe human PEM, and it has been suggested that this phenomenon contributes toward the immune impairments of PEM (Ingenbleek, 1986). T_3 intervention studies have been initiated, therefore, in the authors' laboratory. A low food intake, weanling mouse system has been used which consistently reproduces the serum thyroid hormone response characteristic of human PEM (Filteau and Woodward, 1987a). In this experimental system dietary T_3 improved, many-fold, both a primary thymus-dependent antibody response (Filteau et al., 1987b) and a primary thymus-independent antibody response (Filteau et al., 1987a). The experimental design involved T_3 supplementation throughout the period of underfeeding so that the question remains as to the effectiveness of the hormone when withheld until malnutrition and immune impairment have become established. In a preliminary study, however, T_3 improved a thymus-dependent antibody response when first administered to wasting, immunodepressed animals at the time of primary immunization (Woods and Woodward, 1988). T_3 induces an increase both in total nucleated splenocyte numbers and in the relative number of specifically responding splenic plasma cells (Filteau and Woodward, 1987b; Filteau et al., 1987a, 1987b), but can be effective without increasing total splenocyte numbers (Perry et al., 1988; Woods and Woodward, 1988). Mechanistically, no influence of T_3 is discernible by surface marker analysis of major splenic lymphocyte subpopulations (Filteau, 1987; Filteau and Woodward, 1988a), but functional analyses of such populations are needed with regard to this question. Moreover, the splenic immunoenhancement seen in vivo was not apparent in vitro when examined with regard to a thymus-independent antibody response (Filteau and Woodward, 1988b). Consequently, the effect of T_3 on the spleen appears to be mediated indirectly through one or more alternative organ systems.

At present the scope of immune functions which can be improved by T_3 supplements in the authors' experimental system is not clear. Moreover, dose-response relationships have not been investigated in this program. Consequently, it is unknown whether the improvements effected in antibody responses are dependent on the above-normal blood T_3 levels (total and free) induced in hormone-supplemented malnourished animals. Importantly, however, immunostimulation was achieved in spite of (and independently of) continued wasting malnutrition as assessed by weight change and carcass composition. Dissociation of immune capacity from the wasting condition is demonstrated most clearly in a study on PEM induced in weanling mice by means of a low protein diet (Perry et al., 1988). In this work, T_3-supplemented animals given the imbalanced diet exhibited a greater degree of wasting than their unsupplemented counterparts but, nevertheless, generated a superior primary T-dependent antibody response. It is remarkable that administration of a single compound (T_3) could so influence immune functions in animals continuing to suffer systemic, wasting disease.

The results presumably reflect the power of endocrine hormones to determine metabolic priorities among organ systems, in this case in favor of at least some components of the immune system. Large T_3 supplements impose an added burden on the fat and muscle protein reserves in malnutrition (Carter et al., 1975; Burman et al., 1979; Perry et al., 1988) whereas smaller dosages, within the physiological range, may exert little impact in the short term (Pasquali et al., 1984). Any therapeutic application of T_3 in PEM, therefore, will require improved knowledge of dose-response relationships and care with regard to nutritional assessment and support. On the other hand, a family of thyroid hormone receptors may exist with differing tissue- and gene-activating specificities (Benbrook and Pfahl, 1987; Thompson et al., 1987). It may be possible, therefore, to devise thyromimetic drugs which influence immune functions without affecting whole-body metabolic rate. Organ-specific thyromimetic agents have already been developed (Underwood et al., 1986).

4.2   Glucocorticoids

PEM induces elevated blood levels of total and free glucocorticoid hormones (Becker, 1983). The inhibitory actions of glucocorticoids on immune and inflammatory reactions are widely recognized (Parrillo and Fauci, 1979), and form the basis of the idea that elevated serum levels of these hormones contribute significantly to immunodepression in PEM (Watson, 1984).

Indications that intervention to reduce glucocorticoid levels might be beneficial to immunity in PEM derive from studies with adrenalectomized rodents. Adrenalectomy prior to provision of a low protein diet prevented the lymphopenia observed in intact rats fed the same diet (Aschkenasy, 1957). Similarly, the same surgical procedure reduced the impact of a low protein diet on lymphoid organ weight (presumably lymphocyte numbers) in young mice (Bell et al., 1976). To date, however, no information exists as to the influence of such an intervention on immune functions or disease resistance in experimental PEM.

Work should be done with anti-glucocorticoid drugs. A number of effective inhibitors of glucocorticoid synthesis or activity now exist (Gagne et al., 1985). Their use is clearly of greater clinical relevance than adrenalectomy and would remove the confounding problems of adrenal medullary ablation and elimination of aldosterone. It may also be important to study species which, unlike rodents but like human beings, are glucocorticoid-resistant, i.e., species such as the guinea pig, which has thymocytes that are resistant to lysis by pharmacological concentrations of glucocorticoids (Parillo and Fauci, 1979). Even in the so-called "resistant" species, however, glucocorticoids inhibit mitosis of thymocytes and stimulate mature lymphocytes (Parillo and Fauci, 1979). Anti-glucocorticoid strategies in PEM would be expected to exhibit the disadvantage of reducing the mobilization of substrates from peripheral sites for support of visceral and nervous functions (cf. Becker, 1983). Such a therapeutic approach, therefore, would require adjunct treatments, presumably including nutritional support.

5. Immunostimulatory Drugs

This section is concerned with immunoactive agents which do not occur naturally in vertebrates. Knowledge of such compounds is expanding rapidly (Specter and Hadden, 1985; Werner et al., 1986), but definitive information concerning their usefulness in malnutrition-associated immunodepression is limited.

Lentinan is a polysaccharide derived from the mushroom *Lentinus edodes* (Chihara et al., 1970) and is reported to potentiate anti-tumor immunity (Chihara et al., 1969, 1970) as well as to activate C3, the third component of complement (Hamuro et al., 1978). Sakamoto et al. (1983) investigated the effectiveness of intraperitoneal injections of lentinan in young rats subjected to wasting PEM. This polysaccharide increased serum C3 concentration and total hemolytic complement bioactivity in the malnourished rats, and improved their resistance (measured as survival) to intravenous exposure to living *Listeria monocytogenes*, a facultative intracellular parasite. Importantly, the lentinan injections were effective when given following development of PEM, and did not influence the progress of wasting in the protein-deficient animals as judged by weight change and hematological parameters. This investigation provides the best evidence currently available as to the potential of immunostimulant drugs in the treatment of PEM. Clearly, this area deserves an intensified research effort both in breadth and in depth.

Levamisole is a small molecular weight phenylimidazo derivative of thiazole with immunopotentiating activities which have been exploited in animals and human beings exhibiting a variety of immunodeficiency conditions (Specter and Hadden, 1985). This drug has been reported to improve the depressed microbicidal potential (qualitative nitroblue tetrazolium reduction test) of neutrophils from malnourished children when administered twice weekly for two weeks (Mekkawi et al., 1985). Unfortunately, the cited investigation was conducted as an uncontrolled clinical trial, and it is therefore impossible to determine whether levamisole exerted any influence apart from the inevitable effects of concomitant nutritional rehabilitation. In other work designed as a prospective, double-blind, randomized trial, levamisole reduced the incidence of major sepsis (visceral abscess and/or positive blood culture) and the need for antibiotic treatments in preoperative anergic surgical patients (Meakins et al., 1979). All the patients had gastrointestinal diseases and a large fraction had cancers, so it is reasonable to expect that anergy was related to malnutrition in a substantial proportion of the subjects. In the study of Meakins *et al.* (1979) levamisole was effective within days of initial treatment. Frequently, however, weeks or months are required to achieve results with this drug (Specter and Hadden, 1985), a disadvantage which could reduce its usefulness for meeting acute therapeutic needs in severe PEM. Side-effects, including neutropenia, also are significant with this chemical (Specter and Hadden, 1985). At present it would appear reasonable to direct research attention towards other more potent drugs with less distressing side-effects.

6.   Therapeutic Attempts in Advanced Cancer

Cachexia is a common and intractable condition in advanced malignancies (Balducci and Hardy, 1985; Bistrian, 1986). Similarly, immunodepression can be expected in advanced cancer (Shils, 1979; Wood and Watson, 1984). A multifactorial origin of cancer-associated immunodepression is probable (Shils, 1979; Wood and Watson, 1984), but PEM has frequently been implicated in a determining role (Shils, 1979; Wood and Watson, 1984; Roussel, 1986; van Eys, 1986). To the extent that this idea may be true, it is relevant to make brief mention of some recent developments in cancer immunotherapy.

Much interest recently has surrounded the use of IL-2 either alone (Papa *et al.*, 1986; Rosenberg *et al.*, 1987; West *et al.*, 1987) or in combination with lymphokine-activated killer (LAK) cells (Papa *et al.*, 1986; Ottow *et al.*, 1987; Rosenberg *et al.*, 1987) to combat advanced cancers in experimental animals and human beings. LAK cell activity is produced by incubating the recipient's own lymphocytes with IL-2, and has been attributed mainly to stimulation of natural killer cell activity (Herberman *et al.*, 1987), although a dissenting view exists (Ottow *et al.*, 1987). The evidence is, therefore, that leukocytes from animals and human subjects with advanced cancers (and attendant severe PEM) retain the capacity to respond to IL-2 under appropriate conditions *in vitro*. In a penetrating comment on this controversial therapy, Durant (1987) pointed out that the possible superiority of combination therapy (IL-2 plus LAK cells) over IL-2 alone means that the former may, indeed, reflect successful manipulation of immunity in the cancer patient.

These recent developments suggest that at least the natural killer cell, or a similar entity with broad cytocidal activity, can be stimulated in cancer-associated malnutrition. Other work with enkephalins also supports this conclusion (Faith *et al.*, 1987). In addition, investigations with thymic extracts or hormones (Sztein and Goldstein, 1986) and with Thy⁺ Lyt2⁺ "tumor-infiltrating lymphocytes" (Rosenberg *et al.*, 1986) suggest that the T cell system can be stimulated in at least some types of advanced cancer. Interest is also developing with regard to the possible responsiveness of the bone marrow in cancer patients to exogenous colony-stimulating factors (Sachs, 1987). Neutropenia is among the most important predisposing factors for infection in cancer patients (Bodey, 1986). Consequently, research into the colony-stimulating factors may prove relevant both to the treatment of the malignant condition and to the treatment of malignancy-associated infectious diseases. Overall, it appears that heightened activity can be induced in at least some immunological components in advanced cancer despite the depressive influences of cachexia and other factors.

7.   Concluding Remarks: A Conceptual Foundation for Future Research

Rational immunoenhancement therapies in PEM must be based ultimately on an understanding of the mechanisms underlying the immune impairments in this disease. Broadly speaking, two major differing concepts of causation have been put forward. The first, and much the more popular view at present, is that: "The altered state of the immune system in the malnourished individual is not so much affected by the general state of malnutrition as it is by the deficiency of one or more specific nutrients" (Richter, 1982). This view predicts that supplementation with one or more critical micronutrients will improve immune functions in severe PEM. There appears, however, to be a lack of clear-cut experimental support for this important prediction (cf. Section 2). The alternative concept is that immunodepression is part of a systemic disease syndrome in PEM resulting from the neuroendocrine adaptations which develop in this disease. This view predicts that immunoenhancement therapies in PEM must be based on reestablishing a metabolic priority for immune functions, presumably through administering hormones, their agonists or their inhibitors. Such a therapeutic strategy would be expected to work synergistically with other immunoenhancing treatment modalities. Some preliminary experimental support for this viewpoint derives from the direct intervention studies with T_3 discussed in Section 4.1. These studies suggest an important new physiological principle, viz., that immunological capability and the wasting condition can be dissociated from one another in severe PEM. This, in turn, gives rise to the optimistic concept that acute problems of infectious disease resistance can be addressed independently of the chronic problem of malnutrition.

An underlying assumption of this essay is that the immunodepression which occurs in severe PEM is primarily disadvantageous to the malnourished subject. This idea is unproven and is worthy of investigation. Craddock (1978) and Baker (1986) proposed that the glucocorticoid-associated immunosuppression which occurs in numerous traumatic conditions (e.g., in post-surgical patients) may diminish the likelihood of resultant autoimmune disease. It was postulated (Craddock, 1978) that stressful situations result in exposure of self-antigens which are normally hidden from immune surveillance. Wasting malnutrition is a traumatic, hypercortisolemic condition and may reasonably be considered in the context of Craddock's hypothesis. Methods may be needed, therefore, to achieve finely tuned immunomodulation in PEM with enhanced antimicrobial immunity but suppressed autoimmunity. Alternatively, for the purposes of a short-term therapy for the severely malnourished, the benefits of improved resistance to pathogens may outweigh the risk of precipitating autoimmune reactions. In the latter case an important clinical challenge will be to establish criteria for determining when to terminate immunostimulation therapy. Risk-benefit considerations must be expected to assume significance in treatment of a systemic disease such as wasting PEM. Major side-effects are predictable from some of the potential treatment modalities discussed in this article, e.g., IL-2 (Rosenberg et al., 1986; Durant, 1987; Rosenberg et al., 1987; West et al., 1987); T_3 (Carter et al., 1975; Burman et al., 1979); and zinc (Filteau and Woodward, 1986). Innovative and individualized combinations of strategies will be required to achieve significant, cost-effective immunostimulation in the severe forms of PEM.

8.   Acknowledgments

Preparation of this article was supported by a Medical Research Council of Canada Postgraduate Studentship awarded to SMF and by a Natural Sciences and Engineering Research Council of Canada operating grant given to BW.

References

Ahmed, F., and Qadri, S. S., 1985, Effect of supplementation of essential amino acids on immune response in protein-deficient rats, Nutr. Rep. Int. 31:711.

Anonymous, 1983, Which thymic hormone? Lancet 1:309.

Aschkenasy, A., 1957, On the pathogenesis of anemias and leukopenias induced by dietary protein deficiency, Am. J. Clin. Nutr. 5:14

Auger, C., Monier, J. C., Dardenne, M., Pleau, J. M., and Bach, J. F., 1982, Identification of FTS (facteur thymique serique) on thymus ultrathin sections using monoclonal antibodies, *Immunol. Lett.* **5**:213.

Baker, C. C., 1986, Immune mechanisms and host resistance in the trauma patient, *Yale J. Biol. Med.* **59**:387.

Balducci, L., and Hardy, C., 1985, Cancer and malnutrition--a critical interaction: a review, *Am. J. Hematol.* **18**:91

Barbul, A., Rettura, G., Levenson, S. M., and Seifter, E., 1977, Arginine: a thymotropic and wound-healing promoting agent, *Surg. Forum* **38**:101.

Barbul, A., Wasserkrug, H. L., Seifter, E., Rettura, G., Levenson, S. M., and Efron, G., 1980, Immunostimulatory effects of arginine in normal and injured rats, *J. Surg. Res.* **29**:228.

Barbul, A., Rettura, G., Levenson, A. M., and Seifter, E., 1983, Wound healing and thymotropic effects of arginine: a pituitary mechanism of action, *Am. J. Clin. Nutr.* **37**:786.

Becker, D. J., 1983, The endocrine responses to protein calorie malnutrition, *Ann Rev. Nutr.* **3**:187.

Bell, R., and Hoffman-Goetz, L., 1983, Effect of protein deficiency on endogenous pyrogen-mediated acute phase protein responses, *Can. J. Physiol. Pharmacol.* **61**:376.

Bell, R. C., Hoffman-Goetz, L., and Keir, R., 1985, Monocyte factors modulate *in vitro* T-lymphocyte mitogenesis in protein malnutrition, *Clin. Exp. Immunol.* **63**:194.

Bell, R. G., Hazell, L. A., and Price, P., 1976, Influence of dietary protein restriction on immune competence: II. Effect on lymphoid tissue, *Clin Exp. Immunol.* **26**:314.

Benbrook, D., and Pfahl, M., 1987, A novel thyroid hormone receptor encoded by a cDNA clone from a human testis library, *Science* **238**:788.

Beris, P., and Miescher, P. A., 1988, Hematological complications of anti-infectious agents, *Sem. Hematol.* **25**:123

Besedovsky, H. O., del Rey, A. E., and Sorkin, E., 1985, Immune-neuroendocrine interactions, *J. Immunol.* **135**:750s.

Bhaskaram, P., and Sivakumar, B., 1986, Interleukin-1 in malnutrition, *Arch. Dis. Child.* **61**:182.

Bistrian, B. R., 1986, Some practical and theoretic concepts in the nutritional assessment of the cancer patient, *Cancer* **58**:1863.

Bodey, G. P., 1986, Infection in cancer patients: a continuing association, *Am. J. Med.* **81**:11.

Boyd, W., and Sheldon, H., 1984, *Introduction To the Study of Disease*, 9th ed., p. 187, Lea & Feibiger, Philadelphia.

Bradley, S. F., Kluger, M. J., and Kauffman, C. A., 1987, Age and protein malnutrition: effects on the febrile response, *Gerontol.* **33**:99.

Brown, J. L., 1987, Hunger in the U.S., *Scientific American* **256**:37.

Brown, R. E., and Katz, M., 1967, Passive transfer of delayed hypersensitivity reaction to tuberculin in children with protein calorie malnutrition, *J. Pediat.* **70**:126.

Burman, K. D., Wartofsky, L., Dinterman, R. E., Kesler, P., and Wannemacher, R. W., 1979, The effect of T_3 and reverse T_3 administration on muscle protein catabolism during fasting as measured by 3-methylhistidine excretion, *Metabolism* **28**:805.

Carter, W. J., Shakir, K. M., Hodges, S., Faas, F. H., and Wynn, J. O., 1975, Effect of thyroid hormone on metabolic adaptation to fasting, *Metabolism* **24**:1177.

Castillo-Duran, C., Fisberg, M., Valenzuela, A., Egana, J. I., and Vavy, R., 1983, Controlled trial of copper supplementation during the recovery from marasmus, *Am. J. Clin. Nutr.* **37**:898.

Castillo-Duran, C., Heresi, G., Fisberg, M., and Vavy, R., 1987, Controlled trial of zinc supplementation during recovery from malnutrition: effects on growth and immune function, *Am. J. Clin. Nutr.* **45**:602.

Chandra, R. K., 1979, Serum thymic hormone activity in protein-energy malnutrition, *Clin. Exp. Immunol.* **38**:228.

Chandra, R. K., 1983, Nutrition, immunity, and infection: present knowledge and future directions, *Lancet* **1**:688.

Chandra, R. K., and Newberne, P. M., 1977, *Nutrition, Immunity, and Infection: Mechanisms of Interactions*, pp. 194-196, Plenum Press, New York.

Chihara, G., Maeda, Y., Hamuro, J., Sasaki, T., and Fukuoka, T., 1969, Inhibition of mouse sarcoma 180 by polysaccharide from *Lentinus edodes* (Berk.) sing., *Nature* **222**:687.

Chihara, G., Hamuro, J., Maeda, Y,. Arai, Y., and Fukuoka, T., 1970, Fractionation and purification of the polysaccharides with marked anti-tumor activity, especially lentinan from *Lentinus edodes* (Berk.) sing. (an edible mushroom), *Cancer Res.* **30**:2776.

Craddock, C. G., 1978, Corticosteroid-induced lymphopenia, immunosuppression, and body defense, *Ann. Int. Med.* **88**:564.

Cruz, J. R., Torun, B., Keusch, G., and Goldstein, A. L., 1982, Effect of *in vitro* treatment with thymosin on T-lymphocytes during nutritional recuperation, *Annual Report of the Institute of Nutrition of Central America and Panama*, p. 130.

Dinarello, C. A., and Mier, J. W., 1986, Interleukins, *Ann. Rev. Med.* **37**:173.

Dowd, P. S., and Heatley, R. V., 1984, The influence of undernutrition on immunity, *Clin. Sci.* **66**:241.

Drabik, M. D., Schnure, F. C., Mok, K. T., Moldawer, L. L., Dinarello, C. A., Blackburn, G. L., and Bistrian, B. R., 1987, Effect of protein depletion and short-term parenteral refeeding on the host response to interleukin-1 administration, *J. Lab. Clin. Med.* **109**:509.

Durant, J. R., 1987, Immunotherapy of cancer: the end of the beginning? *N. Engl. J. Med.* **16**:939.

Fabris, N., 1981, Influence of thyroid hormones on the immune system, in: *The "Low T_3 Syndrome"* (R. D. Hesch, ed.), pp. 199-207, Academic Press, London and New York.

Fabris, N. Mocchegiani, E., Muzzioli, M., and Piloni, S., 1986, Recovery of age-related decline of thymic endocrine activity and PHA response by lysine-arginine combination, *Int. J. Immunopharmac.* **8**:677.

Faith, R. E., Liang, H. J., Plotnikoff, N. P., Murgo, A. J., and Nimeh, N. F., 1987, Neuroimmunomodulation with enkephalins: *in vitro* enhancement of natural killer cell activity in peripheral blood lymphocytes from cancer patients, *Natural Immun. Cell Growth Regul.* **6**:88.

Filteau, S. M., 1987, The Influence of Supplemental Triiodothyronine on the Immune System of Malnourished Mice, Ph.D. thesis, University of Guelph, Guelph, ON.

Filteau, S. M., and Woodward, B., 1982, The effect of severe protein deficiency on serum zinc concentration of mice fed a requirement level or a very high level of dietary zinc, *J. Nutr.* **112**:1974.

Filteau, S. M., and Woodward, B., 1984, Relationship between serum zinc level and immunocompetence in protein-deficient and well-nourished weanling mice, *Nutr. Res.* **4**:853.

Filteau, S. M., and Woodward, B., 1986, Zinc and immunity, *Nutr. Rev.* **44**:283.

Filteau, S. M., and Woodward, B., 1987a, Influence of severe protein deficiency and of severe food intake restriction on serum levels of thyroid hormones in the weanling mouse. *Nutr. Res.* **7**:101.

Filteau, S. M., and Woodward, B., 1987b, The effect of triiodothyronine on the primary antibody response to sheep red blood cells in moderately undernourished weanling mice, *Nutr. Res.* **7**:755.

Filteau, S. M., and Woodward, B., 1988a, Splenic lymphocyte subpopulations in severe malnutrition with or without triiodothyronine immunostimulation, *Proc. Can. Fed. Biol. Soc.* **31**:99 (Abstr.).

Filteau, S. M., and Woodward, B., 1988b, Physiological and supraphysiological supplements of triiodothyronine do not influence the primary in vitro antibody response to trinitrophenylated Brucella abortus by spleen cells in serum-containing media, *Acta Endocrinol. (Copenh.)* **118**:351.

Filteau, S. M., Berdusco, E., Perry, K. J., and Woodward, B., 1987a, The effect of triiodothyronine on evanescent delayed hypersensitivity to sheep red blood cells and on the primary antibody response to trinitrophenylated *Brucella abortus* in severely undernourished weanling mice, *Int. J. Immunopharmac.* **9**:811.

Filteau, S. M., Perry, K. J., and Woodward, B., 1987b, Triiodothyronine improves the primary antibody response to sheep red blood cells in severely undernourished weanling mice, *Proc. Soc. Exp. Biol. Med.* **185**:427.

Frayn, K. N., 1986, Hormonal control of metabolism in trauma and sepsis, *Clin. Endocrinol.* **24**:577.

Friedman, P. J., 1987, Is wasting itself lethal? A case-control prospective study, *Nutr. Res.* **7**:707.

Fudenberg. H. H., 1985, "Transfer factor": an update, *Proc. Soc. Exp. Biol. Med.* **178**:327.

Gagne, D., Pons, M., and Philibert, D., 1985, RU 38486: a potent antiglucocorticoid in vitro and in vivo, *J. Steroid Biochem.* **3**:247.

Gershwin, M. E., Beach, R. S., and Hurley, L. S., 1985, *Nutrition and Immunity*, Academic Press, Orlando, FL.

Golden, B. E., and Golden, M. H. N., 1981a, Plasma zinc, rate of weight gain, and the energy cost of tissue deposition in children recovering from severe malnutrition on a cow's milk or soya protein based diet, *Am. J. Clin. Nutr.* **34**:892.

Golden, M. H. N., and Golden, B. E., 1981b, Effect of zinc supplementation on the dietary intake, rate of weight gain, and energy cost of tissue deposition in children recovering from severe malnutrition, *Am. J. Clin. Nutr.* **34**:900.

Golden, M. H. N., Jackson, A. A., and Golden, B. E., 1977, Effect of zinc on the thymus of recently malnourished children, *Lancet* **2**:1057.

Golden, M. H. N., Golden, B. E., Harland, P. S. E. G., and Jackson, A. A., 1978, Zinc and immunocompetence in protein-energy malnutrition, *Lancet* **1**:1226.

Goldstein, A. L., Low, T. L. K., Thurman, G. B., Zatz, M. M., Hall, N., Chen, J., Hu, S.-K., Naylor, P. B., and McClure, J. E., 1981, Current status of thymosin and other hormones of the thymus gland, *Recent Prog. Horm. Res.* **37**:369.

Goldstein, G., and Audhya, T. K., 1985, Thymopoietin to thymopentin: Experimental studies, *Surv. Immunol. Res.* **4** (Suppl. 1):1.

Gross, R. L., and Newberne, P. M., 1980, Role of nutrition in immunologic function, *Physiol. Rev.* **60**:188.

Gusberg, R. J., 1986, The multiple trauma victim: nutritional cripple, *Yale J. Bio. Med.* **59**:403.

Halstead, J. A., and Smith, J. C., 1970, Plasma zinc in health and disease, *Lancet* **1**:322.

Hamuro, J., Hadding, U., and Bitter-Svermann, D., 1978, Solid phase activation of alternative pathway of complement by β-1,3-glucans and its possible role for tumor regressing activity, *Immunology* **34**:695.

Hansen, M. A., Fernandes, G., and Good, R. A., 1982, Nutrition and immunity: the influence of diet on autoimmunity and the role of zinc in the immune response, *Ann. Rev. Nutr.* **2**:151.

Herberman, R. B., Hiserodt, J., Vujanovic, N., Balch, C., Lotzova, E., Bolhuis, R., Golub, S., Lanier, L. L., Phillips, J. H., Riccardi, C., Ritz, J., Santoni, A., Schmidt, R. E., and Uchida, A., 1987, Lymphokine-activated killer cell activity: characteristics of effector cells and their progenitors in blood and spleen, *Immunol. Today* **8**:178.

Hoffman-Goetz, L., and Kluger, M. J., 1979a, Protein deficiency: its effects on body temperature in health and disease states, *Am. J. Clin. Nutr.* **32**:1423.

Hoffman-Goetz, L., and Kluger, M. J., 1979b, Protein deprivation: its effects on fever and plasma iron during bacterial infection in rabbits, *J. Physiol.* **295**:419.

Hoffman-Goetz, L., McFarlane, D., Bistrian, B. R., and Blackburn, G. L., 1981, Febrile and plasma iron responses of rabbits injected with endogenous pyrogen from malnourished patients, *Am. J. Clin. Nutr.* **34**:1109.

Hoffman-Goetz, L., Bell, R. C., and Keir, R., 1985, Effect of protein malnutrition and interleukin 1 on *in vitro* rabbit lymphocyte mitogens, *Nutr. Res.* **5**:769.

Hoffman-Goetz, L., Keir, R., and Young, C., 1986, Modulation of cellular immunity in malnutrition: effect of interleukin 1 on suppressor T cell activity, *Clin. Exp. Immunol.* **65**:381.

Ingenbleek, Y., 1986, Thyroid dysfunction in protein-calorie malnutrition, *Nutr. Rev.* **44**:253.

Irving, M., 1985, Enteral and parenteral nutrition, *Br. Med. J.* **291**:1404.

Itoh, T., Kasahara, S., and Mori, T., 1982, A thymic epithelial cell line, IT-45R1, induces the differentiation of prethymic progenitor cells into postthymic cells through direct contact, *Thymus* **4**:69.

Jackson, T. M., and Zaman, S. N., 1980, The *in vitro* effect of the thymic factor thymopoietin on a subpopulation of lymphocytes from severely malnourished children, *Clin. Exp. Immunol.* **39**:717.

Jose, D. G., Ford, G. W., and Welch, J. S., 1976, Therapy with parent's lymphocyte transfer factor in children with infection and malnutrition, *Lancet* **1**:263.

Kauffman, C. A., Jones, P. G., and Kluger, M. J., 1986, Fever and malnutrition: endogenous pyrogen/interleukin-1 in malnourished patients, *Am. J. Clin. Nutr.* **44**:449.

Khalil, M., Kabiel, A., El-Khateeb, S., Aref, K., El Lozy, M., Jahin, S., and Nasr, F., 1974, Plasma and red cell water and elements in protein-calorie malnutrition, *Am. J. Clin. Nutr.* **27**:260.

Koster, F., Gaffar, A., and Jackson, T. M., 1981, Recovery of cellular immune competence during treatment of protein-calorie malnutrition, *Am. J. Clin. Nutr.* **34**:887.

Malavé, I., and Benain, I. R., 1984, Modulatory effect of zinc on the proliferative response of murine spleen cells to polyclonal T cell mitogens, *Cell. Immunol.* **89**:322.

Mansour, M. M., Mikhail, M. M., and Guirgis, N. I., 1983, Effect of zinc supplementation on *S mansoni*-infected hamsters, *Ann. Trop. Med. Parasitol.* **77**:517.

Meakins, J. L., Christou, N. V., Shizgal, H. M., MacLean, L. D., 1979, Therapeutic approaches to anergy in surgical patients, *Ann Surg.* **190**:286.

Mekkawi, T. E., Seoud, H, A., El Saharty, M. M., Ali, G. A., and Mandour, T. M., 1985, The role of immuno restorative agent in malnourished children and infants, *J. Egyptian Soc. Parasitol.* **15**:623.

Mittal, A., and Woodward, B., 1985, Thymic epithelial cells of severely undernourished mice: accumulation of cholesteryl esters and absence of cytoplasmic vacuoles, *Proc. Soc. Exp. Biol. Med.* **178**:385.

Mittal, A., and Woodward, B., 1986, Ultrastructural and morphometric analysis of thymic epithelial secretory vacuoles in severely protein-energy malnourished weanling mice, *Nutr. Res.* **6**:663.

Mittal, A., Woodward, B., and Chandra, R. K., 1988, Involution of thymic epithelium and low serum thymulin bioactivity in weanling mice subjected to severe food intake restriction of severe protein deficiency, *Exp. Molec. Pathol.*, **48**:226.

Moldawer, L. L., Sobrado, J., Blackburn, G. L., and Bistrian, B. R., 1984, A rationale for administering leukocyte endogenous mediator to protein malnourished, hospitalized patients, *J. Theor. Biol.* **106**:119.

Moldawer, L. L., Hamawy, K. J., Bistrian, B. R., Georgieff, M., Drabik, M., Dinarello, C. A., and Blackburn, G. L., 1985, A therapeutic use for interleukin-1 in the protein-depleted animal, *Br. J. Rheumatol.* **24**:220.

Morgan, P. N., Keen, C. L., Calvert, C. C., and Lonnerdal, B., 1988, Effect of varying dietary zinc intake of weanling mouse pups during recovery from early undernutrition on growth, body composition and composition of gain, *J. Nutr.* **118**:690.

Munster, A. M., Loadholt, C. B., and Leary, A. G., 1977, The effect of antibiotics on cell-mediated immunity, *Surgery* **81**:692.

Murray, M. J., and Murray, A. B., 1977, Starvation suppression and refeeding activation of infection: an ecological necessity? *Lancet* **1**:123.

Nordlind, K, 1985, Stimulating effect of zinc chloride on DNA synthesis of human thymocytes, *Int. Arch. Allergy Appl. Immunol.* **77**:461.

Olusi, S. O., Thurman, G. B., and Goldstein, A. L., 1980, Effect of thymosin on T-lymphocyte rosette formation in children with kwashiorkor, *Clin. Immunol. Immunopathol.* **15**:687.

Ottaway, J. H., and Apps, D. K., 1984, *Biochemistry*, 4th ed., p. 213, Balliere Tindall, London.

Ottow, R. T., Steller, E. P., Sugarbaker, P. H., Wesley, R. A., and Rosenberg, S. A., 1987, Immunotherapy of intraperitoneal cancer with interleukin 2 and lymphokine-activated killer cells reduces tumor load and prolongs survival in murine models, *Cell. Immunol.* **104**:366.

Papa, M. Z., Mule, J. J., and Rosenberg, S. A., 1986, Antitumor efficacy of lymphokine-activated killer cells and recombinant interleukin 2 *in vivo*: successful immunotherapy of established pulmonary metastases from weakly immunogenic and nonimmunogenic murine tumors of three distinct histological types, *Cancer Res.* **46**:4973

Parrillo, J. E., and Fauci, A. S., 1979, Mechanisms of glucocorticoid action on immune processes, *Ann. Rev. Pharmacol. Toxicol.* **19**:179.

Pasquali, R., Baraldi., G., Biso, P., Piazzi, S., Patrono, D., Capelli, M., and Melchionda, N., 1984, Effect of physiological doses of triiodothyronine replacement on the hormonal and metabolic adaptation to short-term starvation and to low-calorie diet in obese patients, *Clin. Endocrinol.* **21**:357.

Perry, K. J., Filteau, S. M., and Woodward, B., 1988, Dissociation of immune capacity from nutritional status by triiodothyronine supplements in severe protein deficiency, *FASEB J.* **2**:2609

Petro, T. M., and Watson, R. R., 1982, Resistance to L1210 mouse leukemia cells in moderately protein-malnourished Balb/c mice treated *in vivo* with thymosin fraction V, *Cancer Res.* **42**:2139.

Petro, T. M., and Wess, J. A., 1987, Thymus derived (T) lymphocyte subsets restore the immune responsiveness of Peyer's patch lymphocytes from mice fed a diet reduced in protein, *Nutr. Res.* **7**:935.

Petro, T. M., Chien, G., and Watson, R. R., 1982, Alteration of cell-mediated immunity to *Listeria monocytogenes* in protein-malnourished mice treated with thymosin fraction V, *Infec. Immun.* **37**:601.

Pinchcofsky, G. D., and Kaminski, M. V., Jr., 1985, Increasing malnutrition during hospitalization: documentation by a nutritional screening program, *J. Am. Coll. Nutr.* **4**:471.

Richter, M. A., 1982, *Clinical Immunology: A Physician's Guide*, 2nd ed., p. 204, Williams and Wilkins, Baltimore and London.

Rosenberg, S. A., Spiess, P., and Lafreniere, R., 1986, A new approach to the adoptive immunotherapy of cancer with tumor-infiltrating lymphocytes, *Science* **233**:1318.

Rosenberg, S. A., Lotze, M. T., Muul, L. M., Chang, A. E., Avis, F. P., Leitman, S., Linehan, W. M., Robertson, C. N., Lee, R. E., Rubin, J. T., Seipp, C. A., Simpson, C. G., and White, D. E., 1987, A progress report on the treatment of 157 patients with advanced cancer using lymphokine-activated killer cells and interleukin-2 or high-dose interleukin-2 alone, *N. Engl. J. Med.* **316**:889

Roussel, E., 1986, Immunoregulatory leucocyte subset typing and PHA response in relation to the nutritional state in cancer patients with gastrointestinal neoplasia, *Diagnostic Immunol.* **4**:10.

Sachs, L., 1987, The molecular control of blood cell development, *Science* **238**:1374.

Sakamoto, M., Ishii, S., and Nishioka, K., 1983, Heightened resistance against *Listeria monocytogenes* infection in malnourished rats after lentinan treatment: correlation with C3 levels, *Nutr. Res.* **3**:705.

Salimonu, L. S., Ojo-Amaize, E., Johnson, A. O. K., Laditan, A. A. O., Akinwolere, O. A. O., and Wigzell, H., 1983, Depressed natural killer cell activity in children with protein-calorie malnutrition: II. Correction of the impaired activity after nutritional recovery, *Cell. Immunol.* **82**:210.

Schuurman, H. J., van de Wijngaert, F. P., Delvoye, L., Broekhuizen, R., McClure, J. E., Goldstein, A. L., and Kater, L., 1985, Heterogeneity and age dependency of human thymus reticuloepithelium in production of thymosin components, *Thymus* **7**:13.

Sheng, F. C., Freischlag, J., Backstrom, B., Kelly, D., and Busuttil, R. W., 1987, The effects of *in vivo* antibiotics on neutrophil (PMN) activity in rabbits with peritonitis, *J. Surg. Res.* **43**:239.

Shils, M. E., 1979, Diet and nutrition as modifying factors in tumor development, *Med. Clin. N. Am.* **63**:1027.

Sidransky, H., 1986, Effects of tryptophan on protein synthesis by liver, in: *Nutritional Diseases: Research Directions in Comparative Pathobiology* (D. G. Scarpelli and G. Migaki, eds.), pp. 71-90, Alan R. Liss, New York.

Simmer, K., Khanum, S., Carlsson, L., and Thompson, R. P. H., 1988, Nutritional rehabilitation in Bangladesh--the importance of zinc, *Am. J. Clin. Nutr.* **47**:1036.

Sirisinha, S., Suskind, R., Edelman, R., Charupatana, C., and Olson, R. E., 1973, Complement and C3-proactivator levels in children with protein-calorie malnutrition and effect of dietary treatment, *Lancet* 1:1016.

Solomons, N. W., 1979, On the assessment of zinc and copper nutriture in man, *Am. J. Clin. Nutr.* 32:856.

Specter, S., and Hadden, J. W., 1985, New approaches to immunotherapy: thymomimetic drugs, *Springer Semin. Immunopathol.* 8:375.

Stinnett, J. D., 1983, *Nutrition and the Immune Response*, Preface, CRC Press, Boca Raton, FL.

Sztein, M. B., and Goldstein, A. L., 1986, Thymic hormones--a clinical update, *Springer Semin. Immunopathol.* 9:1.

Tarnawski, A., and Batko, B., 1973, Antibiotics and immune processes, *Lancet* 1:674.

Thompson, C. C., Weinberger, C., Lebo, R., and Evans, R. M., 1987, Identification of a novel thyroid hormone receptor expressed in the mammalian central nervous system, *Science* 237:1610.

Tocco-Bradley, R., and Kluger, M. J., 1984, Zinc concentration and survival in rats infected with *Salmonella typhimurium*, *Infect. Immun.* 45:332.

Torun, B., Keusch, G. T., Flores-Huerta, S., and Cruz, J. R., 1981, Effect of opsonin replacement in the treatment of severe protein malnutrition, Annual Report of the Institute of Nutrition of Central America and Panama, p. 116.

Underwood, A. H., Emmett, J. C., Ellis, D., Flynn, S. B., Leeson, P. D., Benson, G. M., Novelli, R., Pearce, N. J., and Shah, V. P., 1986, A thyromimetic that decreases plasma cholesterol levels without increasing cardiac activity, *Nature* 325:425.

van Eys, J., 1986, The pathophysiology of undernutrition in the child with cancer, *Cancer* 58:1874.

Walker, A. M., Garcia, R., Pate, P., Mata, L. J., and David, J. R., 1975, Transfer factor in the immune deficiency of protein-calorie malnutrition: a controlled study with 32 cases, *Cell Immunol.* 15:372.

Warner, G. L., and Lawrence, D. A., 1986, Stimulation of murine lymphocyte responses by cations, *Cell Immunol.* 101:425.

Warren, P. J., Hansen, J. D. L., and Lehmann, B. H., 1969, The concentration of copper, zinc, and manganese in the liver of African children with marasmus and kwashiorkor, *Proc. Nutr. Soc.* 28:6A

Watson, R. R., 1984, Stress caused by dietary changes: corticosteroid production, a partial explanation for immunosuppression in the malnourished, in: *Nutrition, Disease Resistance and Immune Function* (R. R. Watson, ed.), pp. 273-283, Marcel Dekker, New York and Basel.

Watson, R. R., and Lim, T. S., 1986, Thymosin fraction 5: effects on T cell functions in mice immunosuppressed by severe dietary protein deficiency. *Int. J. Immunopharmac.* 8:545.

Watson, R. R., Chien, G., and Chung, C., 1983, Thymosin treatment: serum corticosterone and lymphocyte mitogenesis in moderately and severely protein-malnourished mice, *J. Nutr.* 113:483.

Werner, G. H., Floc'h, F., Migliore-Samour, D., and Jolles, P., 1986, Immunomodulating peptides, *Experientia* 42:521.

West, W. H., Taver, K. W., Yannelli, J. R., Marshall, G. D., Orr, D. W., Thurman, G. B., and Oldham, R. K., 1987. Constant-infusion recombinant interleukin-2 in adoptive immunotherapy of advanced cancer, *N. Engl. J. Med.* 316:898.

Wood, C., and Watson, R. R., 1984, Interrelationships among nutritional status, cellular immunity, and cancer, in: *Nutrition, Disease Resistance and Immune Function* (R. R. Watson, ed.), pp. 53-60. Marcel Dekker, New York and Basel.

Woods, J., and Woodward, B., 1988. Immunorestoration with triiodothyronine in severely malnourished mice, *Proc. Can. Fed. Biol. Soc.* **31**:99 (Abstr.).

Zatz, M. M., and Goldstein, A. L., 1985, Mechanism of action of thymosin I. Thymosin fraction 5 increases lymphokine production by mature murine T cells responding in a mixed lymphocyte reaction, *J. Immunol.* **134**:1032.

Chapter 3

The Role of Nutrition in the Prevention and Treatment of Hypertension

Pirjo Pietinen and Antti Aro

1. Introduction

Increase in blood pressure with age is a common finding in industrialized societies. High blood pressure increases the risk of cardiovascular complications like stroke and coronary heart disease. Hypertension can be treated with pharmacological and nonpharmacological means. The pharmacological approach has dominated the field of antihypertensive therapy because nonpharmacological methods have been considered either weak or infeasible. However, many nonpharmacological factors, particularly obesity and the intake of alcohol, fats and various minerals have been found to influence blood pressure both in cross-sectional and prospective studies (Kaplan, 1985).

This article reviews the associations between nutritional factors and blood pressure and gives an overview of dietary measures which are feasible in the prevention and treatment of hypertension. Most of the early studies are covered by referring to review articles, and main attention is paid to controlled studies and to factors of greatest interest during recent years.

Pirjo Pietinen and Antti Aro • Department of Epidemiology, National Public Health Institute, Mannerheimintie 166, SF-00280 Helsinki, Finland.

Advances in Nutritional Research, Vol. 8
Edited by Harold H. Draper
Plenum Press, New York, 1990

2.   Obesity and Weight Reduction

2.1   Overweight and the Development of Hypertension

Blood pressure and obesity are directly related at all ages from child-hood to adulthood (Chiang *et al.*, 1969; Kannel *et al.*, 1976; Stamler *et al.*, 1978). Overweight people have an increased risk of developing hypertension (Chiang *et al.*, 1969) and those who gain weight with age seem to have the greatest risk of becoming hypertensive (Levi *et al.*, 1946; Kannel *et al.*, 1976; Hsu *et al.*, 1977). Studies in which indirect and direct methods of measuring blood pressure were compared have shown that the association between obesity and hypertension is a real one and not an artefact resulting from the confounding effect of increased arm circumference of obese subjects (Chiang *et al.*, 1969).

2.2   Weight Reduction

Early studies on the effects of weight reduction on blood pressure (cf. Chiang *et al.*, 1969) generally demonstrated a fall in blood pressure. Many of these findings are, however, open to criticism, particularly in various methodological aspects (Hovell, 1982).

More recently, the effects of weight reduction on hypertension have been assessed in several controlled studies with random allocation of subjects into treatment groups (Table I). Most of these studies have indicated that weight reduction is beneficial. Reisin *et al.* (1978) observed a significant reduction of systolic and diastolic blood pressure with a decrease in body weight in patients both with and without drug therapy for hypertension. About one-half of the subjects still showed blood pressure levels of 140/90 mm Hg or lower 12-18 months after the start of the weight reduction program (Reisin and Frolich, 1982). Significant reductions of blood pressure have consistently been observed during very low energy diets (320 kcal/d) resulting in weight reductions of more than 20 kg (Tuck *et al.*, 1981; Maxwell *et al.*, 1984). During long-term follow-up the changes in blood pressure have correlated with the changes in body weight (Dornfeld *et al.*, 1985). On the other hand, it is not necessary to achieve normal weight in order to attain a normal blood pressure level (Eliahou *et al.*, 1981).

MacMahon *et al.* (1985) found that weight reduction reduced blood pressure more than placebo and that its effect was equal to, or greater than, that of metoprolol. Other studies have shown that successful weight reduction reduces the need for antihypertensive drug therapy (Ramsay *et al.*, 1978; Imai *et al.*, 1986). In 72% of overweight hypertensive subjects who had been adequately treated with antihypertensive drugs, blood pressure could be successfully controlled by weight reduction after discontinuation of drug therapy (Langford et al., 1985). The mean loss of body weight was 4.5 kg and the duration of follow-up exceeded one year.

Two controlled studies have failed to show a significant effect of weight reduction on blood pressure. Fagerberg et al. (1984) found a slight reduction of blood pressure in a group of 15 men which, however, was statistically nonsignificant. Haynes et al. (1984) studied the effects of behavioral techniques for weight reduction. The program lasted for 16 weeks and the results were assessed six months after the beginning of the study. The degree of weight reduction was rather small and no significant effects on blood pressure were observed.

2.3 Mechanism of Blood Pressure Lowering

The role of sodium in the blood pressure reduction achieved through weight reduction is controversial. Weight reducing diets may have a reduced sodium content, and it has been claimed that the effect of weight reduction on blood pressure is due to a reduction in sodium intake (Dahl et al., 1958). These results are partially supported by the findings of Fagerberg et al. (1984), who found a significant reduction of both systolic and diastolic blood pressure in 15 men treated with weight reduction combined with sodium restriction, while the blood pressure changes were smaller and statistically nonsignificant in another group of 15 men treated with weight reduction without change in sodium intake. The subjects in the first group showed a significant reduction of blood pressure during a preceding 4-week period on a regular diet and without weight reduction (Table I). Changes in alcohol consumption may account for some of the differences observed in this study.

On the other hand, a blood pressure-lowering effect of weight reduction has been documented in several studies, irrespective of changes in sodium intake (Table I). In the study by Reisin et al. (1978), there was no difference in urinary sodium excretion between the groups at the end of the study period. However, sodium intake during the early stages of weight loss was not measured. Tuck et al. (1981) found no difference in the blood pressure response to weight reduction at two levels of sodium intake (40 mmol/d and 140 mmol/d). In the similar study of Maxwell et al. (1984), the lower sodium intake produced a slightly greater response during the first week but there was no difference between the blood pressure responses thereafter. In agreement with these observations, Langford et al. (1985) found that weight reduction reduced blood pressure even in the absence of a simultaneous change in sodium intake and that weight reduction was more efficient than salt restriction in reducing the blood pressure of obese subjects.

Table I. Controlled Studies on the Effect of Weight Reduction on Blood Pressure

Study	Subjects	Diet (Kcal/d), other intervention	Duration	Na excretion mmol/24 h	Change of body weight (kg)	Blood pressure (mm Hg) Systolic/diastolic Initial	Change
Reisin et al. (1978)	24	800–1200, no drugs	4 mo	165	-8.8	157/106	-26[b]/-20[b]
	57	800–1200, drugs	4 mo	185	-9.8	172/113	-37[b]/-23[b]
	26	regular, drugs	4 mo	155	-0.7	171/109	-7/-3
Tuck et al. (1981)	15	320	12 wk	120[a]	-20.2	100[c]	-17[b]
	10	320	12 wk	40[a]	-20.2	105[c]	-19[b]
Fagerberg et al. (1984)	15	1220	12 wk	195	-8.7	152/99	-3/-4
	15	regular	4 wk	177	-0.1	156/101	-10[b]/-4[b]
		1220	9 wk	96	-8.2	146/98	-7[b]/-7[b]
Maxwell et al. (1984)	18	320	12 wk	39	-28	152/100	-26[b]/-22[b]
	12	320	12 wk	200	-26	144/95	-29[b]/-24[b]
Haynes et al. (1984)	27	reduced (behavior-oriented)	16 wk	122	-4.1[d]	135/91[d]	+5/+1
	24	regular	16 wk	136	-0.8[d]	134/89[d]	0/0
MacMahon et al. (1985)	20	reduced (-1000 kcal)	25 wk	not determined	-7.4	150/101	-13[e]/-10[ef]
	18	regular, Metoprolol	25 wk	not determined	+2.0	151/101	-10/-6
	18	regular, placebo	25 wk	not determined	+0.5	150/99	-7/-3

[a] Na intake (mmol/24 h)   [b] $P<0.05$ vs. initial   [c] Mean arterial pressure   [d] Six months after start

[e] $P<0.001$ vs. placebo   [f] $P<0.05$ vs. Metoprolol

The hypotensive response to weight reduction is associated with a reduction in circulating blood volume (Reisin *et al.*, 1983), in intracellular water content (Raison *et al.*, 1983), in plasma renin activity and aldosterone concentration (Tuck *et al.*, 1981) and in plasma noradrenaline concentration (Sowers *et al.*, 1982); Reisin *et al.*, 1983; Fagerberg *et al.*, 1984). Hypertensive subjects often have glucose intolerance and increased plasma insulin levels for the degree of overweight, and weight reduction has a favorable effect on these abnormalities. Hyperinsulinemia may stimulate the sympathetic nervous system and increase renal sodium reabsorption, thus contributing to the development of hypertension (Reaven and Hoffman, 1987). Apparently, many hormonal and metabolic factors may influence blood pressure in obese individuals during weight reduction.

2.4   Comments

The majority of studies indicate a definite blood pressure-lowering effect of successful weight reduction. The most impressive results have been observed in studies in which rapid and pronounced weight loss was achieved by very low energy diets (Tuck *et al.*, 1981; Maxwell *et al.*, 1984). In the Framingham Study, a change in relative body weight of 10% resulted in a change of systolic blood pressure, in the same direction, of 6.6 mm Hg in men and of 4.5 mm Hg in women (Ashley and Kannel, 1974). Even minor changes in body weight in near-normal weight patients may have beneficial effects on blood pressure (Imai *et al.*, 1986). Although salt restriction is not necessary for the effect of weight reduction, some results suggest that the best response is achieved by combining reduced salt intake with weight reduction (Gillum *et al.*, 1983; Fagerberg *et al.*, 1984).

3.   Alcohol

3.1   Population Studies

A positive association between alcohol consumption and blood pressure has been documented in some 20 population studies (cf. Criqui, 1987). Although the association has been consistent in most studies, differences between subgroups have been observed. The large Kaiser-Permanente study, comprising 87,000 subjects examined during 1964-68 (Klatsky *et al.*, 1977) indicated an increase in the risk of hypertension from consumption of more than 2-3 drinks per day. In white people consuming more than 6 drinks per day the risk was doubled, whereas in black people the excess risk was smaller.   A more recent series of observations from the same health care program (Klatsky *et al.*, 1986), covering more than 66,000 people without drug therapy for hypertension, confirmed the relationship and indicated that it was stronger in men, in white people, and in persons older than 55 years of age. In men the relationship was continuous, starting from 1-2 drinks per day, but in women the increase in blood pressure was observed only at three or more drinks per day.

The results of the Lipid Research Clinics (LRC) prevalence study (Criqui *et al.*, 1981) also showed a more consistent relationship between alcohol consumption and blood pressure in men than in women. Women showed a J-shaped relationship with lowest blood pressure levels at 2-3 drinks per day. Whether there is a threshold in the blood pressure effect of alcohol is controversial. Among studies assessing the dose response of alcohol, about one-half have indicated a threshold while the others have shown a more or less linear relationship (Criqui, 1987). After adjustment for confounding factors like age, obesity, and smoking, the J-shape was reduced or abolished in many studies (Cooke *et al.*, 1983; MacMahon *et al.*, 1984; Criqui, 1987). It has been estimated that each daily alcoholic drink increases systolic blood pressure by 0.9 mm Hg in men (Elliott *et al.*, 1987). Alcohol consumption reduces the efficacy of antihypertensive drug therapy (Puddey *et al.*, 1987). Part of this effect may be explained by reduced compliance with drug treatment in heavy alcohol consumers (Lang *et al.*, 1987). In alcoholics, hypertension is at least twice as common as in the general population (Cruickshank *et al.*, 1985).

3.2   Mode of Action

An acute pressor effect of alcohol has been demonstrated in many studies (Grollman, 1930; Orlando, 1976; Kupari, 1983; Potter *et al.*, 1986) but not in all subjects (Puddey *et al.*, 1985b; Beevers *et al.*, 1986). It seems to be independent of sympathetic activity (Kupari *et al.*, 1983; Potter *et al.*, 1986) and may be connected to calcium ion fluxes in smooth muscle cells (Potter *et al.*, 1986).

In hypertensive alcohol consumers, abstinence from alcohol reduces both systolic and diastolic blood pressure within a few days (Potter and Beevers, 1984; Malhotra *et al.*, 1985), and in normotensive individuals reduced alcohol consumption leads to a reduction in blood pressure which is proportional to the change in alcohol intake (Puddey *et al.*, 1985a). In the LRC study, blood pressure was more closely correlated with alcohol consumption during the preceding 24 hr than to earlier alcohol intake (Criqui, 1987). These findings are in accordance with the concept of a relatively rapid, direct effect of alcohol on blood pressure. However, as blood pressure recordings are usually performed in the morning, after an overnight abstinence from alcohol, the influence of short, repeated abstinence periods has not been excluded. A significant correlation has been found between blood pressure and the severity of alcohol-withdrawal symptoms (Saunders *et al.*, 1981). During the withdrawal phase, increased urinary epinephrine, plasma norepinephrine and plasma arginine vasopressin levels and plasma renin activity have been observed (Criqui, 1987).

3.3   Comments

There is a strong statistical epidemiological association between alcohol consumption and blood pressure which is more evident for systolic than for diastolic pressure. The relationship is more or less linear for men and J-shaped for women. The association is probably the result of a direct hypertensive effect of alcohol. In some studies estimates have been made of the relative importance of alcohol consumption among the multiple causes of hypertension. These estimates have varied from 5-12% (Friedman *et al.*, 1983; Beevers *et al.*, 1986; Lang *et al.*, 1987) to as high as 20-30% (Cooke *et al.*, 1982; Mathews, 1976). As alcohol apparently is one of the most common treatable causes of hypertension, and as alcohol-related blood pressure seems to lead to the same complications as other forms of hypertension (Friedman *et al.*, 1983; Criqui, 1987) reduction of alcohol consumption should always be attempted in the treatment of hypertension.

4.   Sodium

4.1   Population Studies

The collective results of studies describing sodium intake and blood pressure levels in different populations show an association between systolic and diastolic blood pressure and sodium intake or excretion in both men and women (Meneely and Dahl, 1961; Gleiberman, 1973; Froment, 1979; Simpson, 1985). However, these investigations represent a heterogeneous collection of study designs and methodologies, as well as populations that differ in several other respects in addition to sodium intake.

A large and thoroughly standardized interpopulation study, the INTERSALT project, recently has been conducted in 52 population groups in 32 countries representing a wide range of sodium intakes (INTERSALT, 1988). The study used standardized techniques for blood pressure measurement, collection of 24-hr and spot urine samples, and recorded the major confounding variables. Altogether 10,079 men and women aged 20 to 59 years were studied. Electrolyte-blood pressure relationships were studied in individuals within each center, and the results of regression analyses were pooled for all centers. Relationships between population median electrolyte values and population blood pressure values also were analysed across the centers. Sodium was found to be significantly related to the slope of blood pressure with age in cross-center analyses, but not to median blood pressure or prevalence of high blood pressure. If the relationship were causal, then with 100 mmol lower daily sodium intake (e.g., 70 instead of 170 mmol/day), the average increase in blood pressure from age 25 to age 55 years would be less by 9.0 mm Hg for systolic pressure and 4.5 mm Hg for diastolic pressure.

Contrary to the findings from between-population comparisons, most within-population studies have failed to document a relationship between blood pressure and sodium intake or excretion (cf., Harlan and Harlan, 1986; NIH, 1986). In the INTERSALT study, sodium excretion was found to be significantly related to blood pressure in within-center analyses, and at least in part this relationship was independent of body mass index and alcohol intake. It was estimated that a 100 mmol/day lower sodium intake corresponded statistically to an average blood pressure reduction of 3.5 mm Hg (systolic) and 1.5 mm Hg (diastolic) when adjusted for age and sex. However, further adjustment for potassium, body mass index and alcohol reduced these estimates to 2.2 mm Hg (systolic) and 0.1 mm Hg (diastolic).

In the INTERSALT data there is a large gap of 50 mmol/day between the median sodium intake in the four low sodium populations and the next lowest intake. It is possible that the relationship between sodium intake and blood pressure is weak in the mid-range of sodium intakes and stronger at the lower and upper extremes of intake reported in various populations (Harlan and Harlan, 1986). This is supported by two earlier studies. A positive association between sodium and blood pressure was found in a Japanese town with generally high sodium intake (Kihara et al., 1984), and also among urban Bantu of Zaire, where the mean sodium excretion level is below 100 mmol (M'Buyamba-Kabangu et al., 1986).

4.2   Sodium Restriction

A marked reduction in sodium intake has long been known to significantly decrease blood pressure (Chapman and Gibbons, 1949). However, since a diet containing less than 20 mmol sodium per day is unpalatable, and since only about half of hypertensive patients benefit from it, it is not a real alternative to drug treatment (cf. Laragh and Pecker, 1983). Very rigid sodium restriction may also stimulate renin-angiotensin and the sympathetic nervous system, and in these ways limit both the antihypertensive and potassium-sparing effects seen in moderate sodium restriction (Kaplan, 1985).

The effect of modest sodium restriction has been investigated in many studies since the 1970's (Kaplan, 1985). However, many of these studies have been poorly controlled. This discussion limits the data to controlled studies, which are summarized in Table II in the order of the initial blood pressures of the subjects. These studies involved only untreated hypertensive or normotensive subjects.

Initial blood pressure seems to be the most important factor related to the effect of sodium restriction on blood pressure. Of the five studies on mildly hypertensive subjects (initial blood pressure at least 95 mm Hg), two showed a significant reduction in both systolic and diastolic blood pressure (MacGregor et al., 1982; Chalmers et al., 1986), one only in systolic blood pressure (Parijs et al., 1973), one only in diastolic blood pressure (Morgan et al., 1978), and in the fifth no change in either (Silman et al., 1983). The study of Silman et al. (1983) was, however, exceptional because the control group underwent a remarkable reduction in blood pressure. None of the other studies, including those on normotensive or mild hypertensive subjects, have shown a significant change in blood pressure during salt restriction. When the results for mild hypertensive subjects in the study by Puska et al. (1983) were analyzed separately, no differences in blood pressure change were found between the low-salt and control groups, even among subjects whose initial diastolic blood pressure was 98 mm Hg.

Several studies have indicated that moderate sodium restriction enhances the effect of diuretics (Parijs et al., 1973; Ram et al., 1982; Beard et al., 1982; Carney et al., 1984), as well as beta-blocker treatment (Owens and Brackett, 1978; Erwteman et al., 1984). The sodium intake achieved in these studies has varied from about 40 to 90 mmol. However, in studies in which the hypertensive patients were not able to reduce their sodium intake below 100 mmol, the change was too modest to have any added benefit over drug treatment (Bulpitt et al., 1984; Rissanen et al., 1985).

So far only one study has provided evidence that sodium restriction may prevent the rise in blood pressure in children. Hofman et al. (1983) randomly allocated 476 newborn babies to two groups, one of which had a normal sodium intake and the other a low intake from a low-sodium baby-food formula. The study demonstrated a progressively increasing difference in systolic blood pressure between the two groups during the first six months, after which the study was discontinued.

Table II. Controlled Studies on the Effect of Sodium Restriction on Blood Pressure

Study	Subjects	Study design	Duration	Na excretion mmol/24 h	Blood pressure (mm Hg) Systolic/Diastolic Initial	Change
Parijs et al. (1973)	18 HT	Open crossover	4 wk	L 93 / H 191	175/112	-7^a/+4
MacGregor et al. (1982)	19 HT	Double-blind crossover	4 wk	L 86 / H 162	156/98	-12^a/-6^a, -2/-1
Morgan et al. (1978)	62 HT	Open parallel	2 yr	L 157 / H 191	160/97, 165/97	-5^b/-7^b, -3^b/+2
Silman et al. (1983)	18 HT	Open parallel	24 wk	L 117 / H 159	165/98, 160/98	-26^b/-17^b, -21^b/-11^b
Chalmers et al. (1986)	103 HT	Double-blind parallel	12 wk	L 86 / H 156	153/96, 153/96	-9^a/-6^a, +4/+2
Richards et al. (1983)	12 HT	Open crossover	4–6 wk	L 80 / H 180	150/92	-4/-3
Watt et al. (1983)	18 HT	Double-blind crossover	4 wk	L 87 / H 143	150/91	-14^b/-9^b, -13^b/-8^b
Puska et al. (1983)	72 HT and NT	Open parallel	6 wk	L 77 / H 167	139/90, 138/89	-2/-3^b, -2/-3^b
	15 HT	From above study	6 wk	L – / H –	148/98, 146/97	-1/-4^b, -2/-4^b
Grobbee et al. (1987)	40 NT	Double-blind crossover	6 wk	L 57 / H 129	143/78	-7/-5, -6/-5

Skrabal et al. (1981)	20 NT	Open crossover	2 wk	L 40 H 120	125/73	-3/-3
Watt et al. (1985)	31 NT	Double-blind crossover	4 wk	L 68 H 128	109/62	+1/+3 +2/+2
	35 NT	Double-blind crossover	4 wk	L 56 H 131	113/65	-1/0 1/-2
Cooper et al. (1984)	113 NT (16 yr)	Open crossover	24 d	L 40 H 120	109/62	-7/-2 -5/+3

HT = hypertensive   NT = normotensive   L = low-Na diet   H = high-Na diet

[a] $P < 0.05$ vs. high-Na diet

[b] $P < 0.05$ vs. initial

4.3   Mechanism of Blood Pressure Lowering

Several hypotheses have been developed to connect sodium intake to blood pressure regulation. Of these, the autoregulation theory, the natriuretic hormone hypothesis and altered cellular membrane function have become the most important theoretical concepts. According to the autoregulation theory, the primary abnormality in hypertension is a renal defect leading to sodium and water retention, increased venous return and elevated cardiac output (Guyton et al., 1971). Kidney cross-transplantation experiments in inherited hypertension in rats, as well as kidney transplantation in humans, suggest that the kidney is responsible for the rise in blood pressure in essential hypertension (cf. MacGregor, 1985). According to the natriuretic hormone theory, those who develop primary hypertension have an inherited reduced ability to excrete sodium, which results in a slight increase in blood volume and stimulation of natriuretic hormone release. The combined effects of the natriuretic factor on the kidney and on smooth muscle cells result in inhibition of Na-K-ATPase, leading to natriuresis in the kidney, increased contractile tone in the arteriolar wall, and reduced transcellular membrane sodium-potassium pump activity. According to the third hypothesis, alterations in cell membrane function play a part in the development of primary hypertension (Swales, 1982; Postnow and Orlov, 1984). These alterations in the cell membrane may affect a variety of functions such as cation transport, passive electrolyte movement, and intra- and extracellular calcium binding. The final consequence would be raised intracellular calcium levels leading to increased peripheral resistance.

4.4.   Comments

Based on the data from randomized, controlled trials, moderate sodium restriction may lower the blood pressure only of those hypertensive subjects whose habitual level of diastolic blood pressure is over 95 mm Hg. Old people and blacks seem to benefit from sodium restriction more than young people.

Genetic factors have been implicated in the response of blood pressure to sodium restriction. The hypothesis that people with a family history of high blood pressure are more responsive to dietary sodium has not been corroborated in controlled studies (Watt et al., 1985). The value of plasma renin in discriminating sodium-sensitive and non-sensitive hypertensives needs further consideration.

At the moment no methods are available to detect sodium-sensitive persons. However, there is no apparent harm in moderately restricting the sodium intake of hypertensive subjects to 75-100 mmol/day. This level of intake is achievable by deleting most high-sodium foods from the diet and adding no extra salt in cooking or at the table.

5. Potassium

5.1   Population Studies

An inverse correlation between potassium excretion and blood pressure has been found in most, but not all, studies within populations in various parts of the world (cf. NIH, 1986). In the INTERSALT study, potassium was found to be negatively correlated with blood pressure in individuals but across centers there was no consistent association (INTERSALT, 1988). Froment and coworkers (1979) were the first to show a positive, high-order correlation between the urinary sodium/potassium ratio and blood pressure between populations. The sodium/potassium ratio has also been found to be positively correlated with blood pressure in numerous within-population studies (cf. Harlan and Harlan, 1986; NIH, 1986). This finding was confirmed in the INTERSALT study.

5.2   Potassium Supplementation

Increased dietary potassium was claimed to lower blood pressure as long as 60 years ago (Addison, 1928). However, most studies carried out up to now have been poorly controlled. The results of more recent, better controlled trials are summarized in Table III. The studies are ranked in descending order of initial blood pressure. Significant reductions were reported in six of the eleven trials. Of the eight trials on mild to moderate untreated hypertensives, four showed a significant blood pressure reduction (Iimura et al., 1981; McGregor et al., 1982; Matlou et al., 1986; Siani et al., 1987), two showed no effect (Richards et al., 1984; Zoccali et al., 1985) and in two the results were uncertain (Overlack et al., 1983; Svetkey et al., 1987). The significant difference in blood pressure initially found by Svetkey et al. (1987) was diminished by correction for the higher baseline blood pressure of the intervention group. In the study by Zoccali et al. (1985) potassium supplementation was provided using a salt substitute and most of the subjects complained about its taste. Of the three studies carried out among normotensive subjects (Burstyn et al., 1980; Khane and Thom, 1982; Barden et al., 1986), only one showed a reduction in blood pressure (Khane and Thom, 1982).

Table III. Controlled Studies on the Effect of Weight Reduction on Blood Pressure

Study	Subjects	Study design	Duration	24-h excretion (mmol)			Blood pressure (mm Hg) Systolic/Diastolic	
				Initial Na	K	Changed K	Initial	Change
Iimura et al. (1981)	20 HT	Open crossover; 25 or 175 mmol	10 d	158	41	124	114 (MAP)	−8[a]
Matlou et al. (1986)	32 HT	Single-blind crossover, 1 tsp KCl-salt (65 mmol) vs. placebo	6 wk	144	51	114	154/105	−7[a]/−3[a]
MacGregor et al. (1982)	23 HT	Double-blind crossover, 60 mmol vs. placebo	4 wk	152	68	118	154/99	−7[a]/−4[a]
Overlack et al. (1983)	16 HT	Open fixed order; 100 mmol vs. no supplement	8 wk	181	66	153	152/98	−17[b]/−10[b] −7/−2
Zoccali et al. (1985)	19 HT	Single-blind crossover, 100 mmol vs. lactose	2 wk	182	58	139	154/96	−2/−2
Svetkey et al. (1987)	101 HT	Double-blind parallel, 120 mmol vs. placebo	8 wk				148/95 142/94	−6[b]/−4[b] 0/−2
Siani et al. (1987)	32 HT	Double-blind crossover, 48 mmol vs. placebo	15 wk	183	57	87	145/92	−14[a]/−10.5[a]

Richards et al. (1984)	12 HT	Open crossover, 140 mmol in flevored elixir vs. control diet	4-6 wk	180^c	60^c	200^c	149/92	$0/-1^d$
Burstyn et al. (1980)	21 NT	Open crossover, 80 mmol vs. control diet	22 d	171	53	117	115/80	0/0
Khaw and Thom (1982)	20 NT	Double-blind crossover, 64 mmol vs. placebo	2 wk	138	73	130	118/74	$-1/-2^a$
Barden et al. (1986)	44 NT	Double-blind crossover, 80 mmol vs. placebo	4 wk	130^c	55^c	110^c	118/71	-1/-1

[a] P<0.05 vs. control   [b] P<0.05 vs. initial   [c] Values taken from figures   [d] Intra-arterial

Controlled studies on the effect of combined sodium restriction and potassium supplementation are summarized in Table IV. Among mild hypertensives, this combination has been shown to reduce blood pressure (Holly et al., 1981). However, it is possible that the reduction in sodium intake was the critical factor. Most studies in which sodium restriction plus potassium supplementation was compared to sodium restriction alone have shown no difference between the effects of these two treatments, either among mild hypertensives or normotensives (Parfrey et al., 1981; Skrabal et al., 1981; Skrabal et al., 1984; Smith et al., 1985). However, the number of subjects in these studies has been small and the time periods have been short (from two to four weeks). The study by Chalmers et al. (1987), comprising 200 mildly hypertensive subjects, was exceptionally large. Either sodium restriction or potassium supplementation, or their combination, significantly reduced both systolic and diastolic blood pressure over 12 weeks compared to that of control subjects.

Potassium supplementation has been shown to reduce the blood pressure of hypertensive subjects treated with diuretics. In a double-blind crossover study on patients who had diuretics-induced hypokalemia, a potassium supplement of 60 mmol per day reduced mean blood pressure by an average of 5.5 mm Hg (Kaplan et al., 1985). However, another study in normokalemic patients showed no blood pressure effect of potassium supplementation (Bulpitt et al., 1985). Further, a combination of potassium supplementation with moderate sodium restriction in hypertensive subjects treated with diuretics did not further improve blood pressure control beyond that achieved with sodium restriction alone (Skrabal et al., 1984).

5.3   Mode of Action

Several mechanisms have been postulated to account for the antihypertensive action of potassium (Treasure and Ploth, 1983). Potassium could act as a diuretic agent, thereby reducing extracellular volume and, in turn, blood pressure. Potassium may alter the activity of the renin-angiotensin system and reduce the influence of angiotensin on vascular, adrenal or renal receptors. There is some evidence that potassium modifies central or peripheral neural mechanisms that regulate blood pressure. In addition, high potassium diets could reduce blood pressure by relaxing vascular smooth muscle, thereby reducing peripheral vascular resistance directly. In some animal models reduction of the Na/K ratio has lowered blood pressure. Further, high potassium intake is effective in reducing blood pressure only in salt-dependent hypertension. These data do not allow a definite conclusion regarding a common mechanism through which potassium exerts these effects.

5.4   Comments

There is growing evidence that, within populations, dietary potassium is negatively associated with blood pressure and that the sodium/potassium ratio is positively associated with blood pressure. A high dietary intake of potassium also has been associated with a reduced risk of stroke-associated mortality (Khaw et al., 1987).

Potassium supplementation does not seem to have an effect on blood pressure among normotensive persons, but may have a modest effect in at least some patients with mild to moderate essential hypertension. The magnitude of the decrease in blood pressure observed in patients receiving potassium supplements in different human studies seems to be related to the concomitant sodium intake: the higher the sodium intake the greater the blood pressure response. Moderate dietary supplements of potassium might be a valuable alternative to pharmacological methods for controlling blood pressure in patients who have less severe hypertension, particularly those who are not willing or able to restrict salt intake. The dramatic decrease in blood pressure in blacks taking potassium supplements shown by Svetkey et al. (1987) suggests that this subgroup may be particularly sensitive to potassium treatment. The possibility of achieving a substantial increase in potassium intake by acceptable, simple dietary modifications should be explored, as well as the long term safety and tolerability of this type of intervention.

6.   Calcium

6.1   Population Studies

The relationship between dietary calcium intake and blood pressure has been analyzed in several studies, most of them conducted in the U.S.A. (cf. Aro, 1987). The results of the first National Health and Nutrition Examination Survey (NHANES-I) conducted in 1971-75 have been analyzed by four independent groups using somewhat different statistical methods. McCarron et al. (1984) concluded that reduced consumption of calcium and potassium were the primary nutritional markers of systolic hypertension in the population studied. Harlan et al. (1984) also found an independent association between dietary calcium and blood pressure. Dietary calcium was inversely related to diastolic blood pressure in women and in black people. In two more recent analyses, based on the same material, the conclusions were negative: no association was found between calcium intake and blood pressure (Gruchow et al., 1985; Sempos et al., 1986).

Table IV. Controlled Studies on the Combined Effect of Sodium Restriction and Potassium Supplementation on Blood Pressure

Study	Subjects	Study design	Duration	24-h excretion (mmol) Na	K	Blood pressure (mm Hg) Systolic/Diastolic Initial	Change
Holly et al. (1981)	15 HT	Observer-blind crossover					
		high-Na	12 wk	230	60	148/96[a]	+6/+6
		low-Na, high-K	12 wk	120	120		-1[b]/-5
	8 NT	As above					
		high-Na	12 wk	252	67	113/63[a]	+7/+3
		low-Na, high-K	12 wk	151	159		+4/+3
Parfrey et al. (1981)	8[e]	Observer-blind crossover					
		low-Na	4 wk	176	82	132/71[a]	-6/-6
		low-Na, high-K	4 wk	164	179		-14/-8
		high-Na	4 wk	304	76		-5/+5
		normal diet	4 wk	186	77		-8/-3
	8[f]	As above					
		low-Na	4 wk	139	79	123/70	-9[c]/-10[c]
		low-Na, high-K	4 wk	154	145		-5/-10[c]
		high-Na	4 wk	282	84		-5/-2
		normal diet	4 wk	143	72		-5/-4
Skrabal et al. (1981)	20 NT	Crossover					
		low-Na	2 wk			125/73	-3/-3
		high-K	2 wk				-2/-4
		low-Na, high-K	2 wk				-2/-3

Reference	N	Design	Duration			BP	Change
Skrabal et al. (1984)	9 HT	Crossover					
		low-Na	4 wk			153/91	-6/-3
		low-Na, high-K	4 wk				0/-3
Smith et al. (1985)	20 HT	Double-blind crossover					
		low-Na run-in	4 wk	68	72	163/103	-3/0
		64 mmol K	4 wk	80	117		-2/0
		vs. placebo	4 wk	73	67		
Chalmers et al. (1987)	200 HT	Double-blind parallel					
		normal diet	12 wk	156	66	153/96	-4/-2^d
		low-Na	12 wk	86	67		-9^d/-6^d
		high-K	12 wk	145	96		-8^d/-5^d
		low-Na, high K	12 wk	73	88		-8^d/-4^d
Grobbee et al. (1987)	40	Double-blind crossover					
		slow-Na	6 wk	129	77	143/78	-6/-5
		placebo	6 wk	57	74		-7/-5
		slow-K	6 wk	69	131		-10^c/-6

^{a}Values taken from figures ^{b}P<0.05 between groups ^{c}P<0.05 compared to initial ^{d}P<0.05 compared to control

^{e}HT parents ^{f}NT parents

Calcium intake has been inversely associated with blood pressure in some other population studies. The association has been strongest in population groups whose mean calcium intakes are relatively low, but it has been evident also in the Netherlands, where the mean calcium intake is 1.0-1.2 g/d (Kromhout *et al.*, 1985; Kok *et al.*, 1986). The most consistent finding in these studies has been an inverse association between the consumption of dairy products and blood pressure. Dairy products contain a variety of nutrients which may affect blood pressure. The high degree of collinearity between the intakes of calcium, potassium and other components of dairy products hampers statistical interpretation of the data (Reed *et al.*, 1985) and, therefore, it is not possible to determine conclusively the independent effects of these nutrients from the results of cross-sectional epidemiological studies. In most studies food intake has been assessed by the 24-hr recall method, which is of limited validity. Elliott *et al.* (1987) could find no association between blood pressure and calcium intake, assessed using seven-day weighed food records, in a group of U.K. men.

6.2   Calcium Supplementation

The effects of dietary calcium supplements in doses of 0.4 to 1.6 g/d on blood pressure have been studied in both hypertensive and normotensive individuals (Table V). In the majority of studies no difference was observed between the effects of calcium and placebo. In three studies on hypertensive subjects (Johnson *et al.*, 1985; MacCarron and Morris, 1985; Strazzullo *et al.*, 1986) systolic blood pressure was reduced during calcium supplementation. In the last two studies the effect was found only in the final measurement. In the first study, all patients were treated with drugs (mainly thiazides, which are known to reduce urinary calcium excretion) throughout the 4-year study period.

In studies on normotensive individuals, diastolic blood pressure was reduced during calcium supplementation in two (Belizan *et al.*, 1983; McCarron and Morris, 1985) and both systolic and diastolic blood pressure in one (Lyle *et al.*, 1987). The individual blood pressure changes were inversely related to basal urinary calcium excretion in the study by Strazzullo et al. (1986). In the study by Grobbee and Hofman (1986), which showed a significant reduction of diastolic blood pressure at 6 weeks but not at 12 weeks, subjects with baseline serum calcium levels below, and serum parathyroid hormone (PTH) concentrations above, the median showed a greater than average reduction of diastolic blood pressure during calcium supplementation.

6.3   Mechanism of Blood Pressure Lowering

Intracellular free calcium influences vascular smooth muscle contraction, and calcium influx-dependent vasoconstriction has been observed in essential hypertension (Robinson, 1984). It is evident that the hypertensive population is heterogeneous with respect to calcium metabolism. Many hypertensive individuals are characterized by lower than average serum ionized calcium concentrations, hypercalciuria and signs of enhanced parathyroid activity (cf. Ljunghall *et al.*, 1987). In certain hypertensive patients increased urinary calcium excretion leads to a reduction in serum ionized calcium concentration and to a secondary increase in PTH secretion (McCarron *et al.*, 1980). Increasing serum ionized calcium concentration by dietary calcium or vitamin D supplements (Ljunghall *et al.*, 1987) may influence blood pressure directly by altering calcium flux into the cells through an effect on calcium binding to cell membranes, or indirectly by decreasing parathyroid activity and thereby decreasing the synthesis of calcitriol.

There are also findings which suggest that the effects of calcium and sodium are interrelated. Calcium supplements lower blood pressure in salt-sensitive subjects but may have the opposite effect in salt-insensitive individuals (Resnick *et al.*, 1986a). In the studies by Resnick *et al.* (1986b), hypertensive subjects with low plasma renin activity were characterized by elevated PTH and calcitriol levels, and an inverse relationship was found between PTH levels and urinary sodium excretion. High sodium intake may be the primary factor leading to increased calcium excretion, decreased serum ionized calcium concentration, and compensatory changes in PTH and calcitriol concentrations (Langford, 1986).

6.4   Comments

In some populations a negative association has been found between the consumption of dairy products and blood pressure. It is so far unknown whether this association is due to calcium or to other constituents of milk products. Calcium supplementation has resulted in a reduction of blood pressure in some studies but in others no effect has been found. Many findings suggest that hypertensive subjects are heterogeneous with respect to the effects of dietary calcium, and that calcium may have a hypotensive effect in a subgroup which is characterized by increased urinary calcium excretion, lower than average serum ionized calcium and increased serum PTH and calcitriol levels. Low plasma renin activity and salt sensitivity may be further characteristics of the calcium-sensitive subpopulation. On the other hand, some hypertensive individuals may react with an increase in blood pressure if their calcium intake is increased. Therefore, it is not

Table V. Controlled Studies on the Effect of Calcium Supplementation on Blood Pressure

Study	Subjects	Study design Dose of Ca	Duration	Blood pressure (mm Hg) Systolic/Diastolic Initial	Change	Notes
Cappuccio et al. (1987)	18 HT	Double-blind, cross-over, 1.6 g/d	1 mo	156/112	Ca +3/+2 P +1/+1	
Strazzullo et al. (1986)	17 HT	Double-blind, cross-over, 1.0 g/d	15 wk	Ca 146/98 P 143/96	-7^{ab}/0 +5/+4	No difference at 3-12 wk Mean Ca intake 0.6 g/d
McCarron and Morris (1985)	48 HT	Double-blind, cross-over, 1.0 g/d	8 wk	152/94	Ca-3^{ab}/0 P 0/0	No difference at 2-6 wk Mean Ca intake 0.8-0.9 g/d
	32 NT		8 wk	121/75	Ca 0/0^{a} P +2/+3	
Nowson and Morgan (1986)	47 HT	Double-blind, parallel 0.4 g/d 0.8 g/d P	2 mo	147/90 156/91 154/91	-5/1 -1/-1 -1/-2	
	48 NT		2 mo	.....	no change	
Bloomfield et al. (1986)	32 HT	Double-blind, parallel 1.5 g/d (n=15) vs. P (n=17)	4 wk	Ca 140/91 P 139/92	+5^{ac}/.. 0^{c}/..	No significant change in diastolic BP, Ca intake <0.75 g/d, 26/32 black patients
Zoccali et al. (1986)	21 HT	Double-blind, cross-over, 1.0 g/d	8 wk	143/89	+3/+2	Change Ca v. P
Johnson et al. (1985)	34 HT	Double-blind, parallel 1.5 g/d (n=18) vs. P (n=16)	4 yr	Ca 141/86 P 140/86	-10^{a}/-2 +4/-2	HT patients with drug therapy, mostly thiazides
	81 NT	1.5 g/d (n=41) vs. P (n=40)	4 yr	Ca 119/73 P 120/75	+5/+5 +4/+3	Mean Ca intake 0.6 g/d

Study	Subjects	Design	Duration	BP	Change	Comments
Grobbee and Hofman (1986)	90 HT	Double-blind, parallel 1.0 g/d (n=46) vs. P (n=44)	12 wk	Ca 143/83 P 143/83	-4/-5 -4/-2	Significant reduction of diastolic BP at 6 wk on Ca. Mean Ca intake 1.3 - 1.4 g/d
Lyle et al. (1987)	75 NT 54 white 21 black	Double-blind, parallel 1.5 g/d (n=37) vs. P (n=38)	12 wk	White Ca 117/72 P 116/73 Black Ca 120/72 P 118/71	-6/-2[d] -2/0[d] -9/+5[d] -4/+5[d]	The differences of systolic and diastolic BP between the Ca and P groups are statistically significant (P<0.02) by repeated measures analysis of covariance. Ca intake <1.0 g/d
van Beresteyn et al. (1986)	58 NT	Double-blind, parallel 1.5 g/d (n=29)	6 wk	Ca 115/66 P 115/65	-6/-2 -4/-3	Mean Ca intake 0.9 -1.0 g/d
Belizan et al. (1983)	57 NT	Double-blind, parallel 1.0 g/d (n=30) vs. P (n=27)	22 wk	Men Ca 118/74 P 109/68 Women Ca 104/69 P 100/68	0/-9[a]% +1/-1% -1[a]/-6[a]% +1/+1%	Mean changes over 7-22 wk in men and 9-22 wk in women. Mean Ca intake 0.6 - 0.8 g/d

HT = hypertensive   NT = normotensive   P = placebo

[a]P<0.05 vs. placebo   [b]P<0.05 vs. initial   [c]Change at 2 wk   [d]Values adjusted for baseline means, age, skinfolds, Ca intake and compliance

surprising that no overall effect of calcium supplementation on blood pressure has been observed in the majority of studies.

7.   Magnesium

There are few epidemiological studies on the possible association between magnesium intake and blood pressure. Urinary magnesium excretion was found to be negatively correlated with blood pressure in Belgian army subjects but not among Koreans (Kesteloot, 1984). In a population study in two Belgian towns no association with 24-hr urinary magnesium excretion was found (Staessen et al., 1983). In dietary surveys, magnesium intake has rarely been analyzed in relation to blood pressure. Of the 61 dietary variables that were analyzed in the Honolulu heart study, magnesium intake had the strongest association with blood pressure (Joffres et al., 1987). However, it was not possible to separate any effect of magnesium from that of other variables because of high correlation between magnesium intake and that of many other nutrients.

Dyckner and Wester (1983) carried out an uncontrolled study to investigate the effects of magnesium supplementation (15 mmol/d as aspartate hydrochloride) in 39 patients receiving long term treatment with diuretics. In six months, a significant reduction in both systolic and diastolic blood pressure occurred, though no changes were found in plasma electrolyte concentrations. A double-blind crossover study on the effect of supplementing 17 previously untreated hypertensive patients with 15 mmol magnesium as aspartate HCl revealed increased plasma and urine magnesium in the magnesium-treated group, but no change in blood pressure after one month of treatment (Cappuccio et al., 1985). Nineteen mmol Mg given as magnesium oxide to eight untreated hypomagnesemic patients with recent onset of high-renin essential hypertension, similarly exerted no effect on blood pressure (Cohen et al., 1984). The effect of magnesium supplementation (about 12 mmol Mg as magnesium oxide) has also been studied in a double-blind, randomized multicenter study on 40 hypertensive patients treated with potassium-depleting diuretics (Henderson et al., 1986). No changes in plasma magnesium level or blood pressure were observed after three or six months.

Experimental evidence supports a major role of magnesium in the control of vascular tone (cf. Wester and Dyckner, 1987). Rats with dietary magnesium deficiency develop high blood pressure in proportion to the severity of the deficiency. Hypomagnesemia increases vascular resistance and blood pressure more effectively in combination with hypokalemia or hypercalcemia than by itself.

There is no evidence that magnesium supplementation alone lowers blood pressure in untreated hypertensive patients. In hypertensive subjects treated with diuretics, supplementation of the diet with magnesium seems to have little or no effect on blood pressure if the plasma concentration is in the normal range. However, hypomagnesemia, which is frequently induced by diuretic therapy, should be avoided and corrected. There is a need for more clinical data on the interactions of magnesium, potassium and calcium. For instance, it has been suggested that lack of potassium might explain the negative results of magnesium supplementation of hypertensive patients given diuretics (Wester and Dyckner, 1987).

8. Dietary Fat

8.1   Population Studies

Cross-sectional epidemiological studies have provided some evidence that dietary fat influences blood pressure regulation in man. In the Western Electric Study, diastolic blood pressure was inversely related to the intake of monounsaturated and polyunsaturated fatty acids, whereas no association was found between blood pressure and the intake of saturated fats (Nichaman et al., 1984). A positive correlation was found between mean arterial pressure and an index of saturated fat intake in a population sample of over 8000 adults in Eastern Finland (Salonen et al., 1983). In contrast, no association between blood pressure and either total saturated or polyunsaturated fats or linoleic acid was found in the NHANES survey (McCarron et al., 1984; Gruchow et al., 1985).

Correlations between blood pressure and the relative linoleic acid content of adipose tissue triglycerides or plasma lipid factions have been examined in four studies. A negative association between adipose tissue linoleic acid concentration and blood pressure was found in a survey among 450 German men (Oster et al., 1979). A negative correlation between blood pressure and the relative linoleic acid content of serum phospholipids and cholesterol esters has been described in two groups of Finnish men (Miettinen et al., 1982; Riemersma et al., 1986). In a population sample of middle-aged American men adipose linoleic acid was not associated with blood pressure, whereas an absolute increase of 1% in linolenic acid was associated with a decrease of 5 mm Hg in systolic, diastolic and composite mean arterial blood pressure (Berry and Hirsch, 1986).

8.2   Clinical Trials

A summary of data from human studies on the effect of dietary fat on blood pressure is presented in Table 6. An effect of dietary fat on blood pressure in humans was first suggested by Iacono *et al.* in 1975. Reduction of dietary fat content from 42% to 20% of energy with a concomitant increase in the P/S ratio from 0.2 to 1.0 lowered both systolic and diastolic blood pressure among normotensive adults within 40 days. Blood pressure remained low when the subjects were transferred to a diet containing 35% of calories from fat with a P/S ratio of 1.0. These effects were not confirmed in a second study on normotensive subjects but were observed in hypertensives (Iacono *et al.*, 1981). A third study carried out by the same group showed that systolic blood pressure was consistently lower during periods when a diet with a P/S ratio of 1.0 as opposed to a ratio of 0.3 was consumed, whereas total fat intake did not appear to affect blood pressure (Judd *et al.*, 1981).

Two other studies support the hypothesis that an increase in the P/S ratio without simultaneous reduction in total fat intake lowers blood pressure (Comberg *et al.*, 1978; Stern *et al.*, 1980). All these studies suggest that the hypotensive effect of fat modification was due to the increased intake of linoleic acid. However, in two experiments in which the subjects were fed either supplemental linoleic or oleic acid, no difference in the effects of these acids was found. In a study on mildly hypertensive subjects both safflower and groundnut oil decreased diastolic blood pressure relative to placebo (Rao *et al.*, 1981), whereas in normotensives neither supplemental linoleic acid nor oleic acid had any effect on blood pressure (Sacks *et al.*, 1987). In two other studies, changing the P/S ratio alone or in combination with a low-fat or high-fat diet did not change the blood pressure of normotensive subjects (Brussaard *et al.*, 1981; Margetts *et al.*, 1984).

The effect of a simultaneous reduction in fat content and increase in P/S ratio of the diet has been studied in a series of clinical trials among middle-aged North Karelians in Finland (Iacono *et al.*, 1983; Puska *et al.*, 1983; Puska *et al.*, 1985). Lowering the fat content from 38-39% to 23-24% of energy and increasing the P/S ratio from 0.2 to 0.9-1.2 reduced blood pressure in six weeks, and increasing the P/S ratio to 0.4 reduced blood pressure in twelve weeks. In all three experiments the values returned to initial levels after switchback to the original diet.

In contrast to the above three studies, the double-blind crossover study by Sacks *et al.* (1987a) on mild hypertensives revealed no effect on blood pressure of replacing dietary saturated fat with either linoleic acid or carbohydrate. Two of the three other double-blind studies also yielded negative results (Margetts *et al.*, 1984; Sacks *et al.*, 1987b).

Table VI. Controlled Studies on the Effect of Dietary Fat on Blood Pressure

Study	Subjects	Study design	Duration wk	Blood pressure (mm Hg) Systolic/Diastolic Initial	Change
Iacono et al. (1975)	21	Fixed order			
		42% fat, P/S 0.22	2	136/80	
		25% fat, P/S 1.0	6		-13[a]/-7[a]
		35% fat, P/S 1.0	6		+1/+3
		42% fat, P/S 0.22	6		+5/+7
Iacono et al. (1981)	10 ET	Fixed order			
		44% fat, P/S 0.3	3	138/98	
		25% fat, P/S 0.98	4		-8[a]/-8[a]
		44% fat, P/S 0.3	4		+6/+3
	10 NT	44% fat, P/S 0.3	3	124/83	
		25% fat, P/S 0.98	4		-2/-2
		44% fat, P/S 0.3	4		+1/0
Comberg et al. (1978)	8	Habitual diet	-	135/92	
		'linoleic acid enriched'	4		-8/-7[a]
		habitual diet	12		+9/+6
Stern et al. (1980)	47	Parallel			
		habitual diet	6	139/80	-5/-1
		PUFA 20% of energy	6	144/82	-11[b]/-2
Rao et al. (1981)	24	Double-blind, parallel			
		placebo	6	151/98	-6/-1[c]
		20 ml groundnut oil	6	150/99	-6/-7[c]
		20 ml safflower oil	6	146/99	-4/-9[c]

Table VI. Continued

Study	Subjects	Study design	Duration wk	Blood pressure (mm Hg) Systolic/Diastolic Initial	Change
Sacks et al. (1987a)	21 HT	Double-blind, crossover			
		33% fat, P/S 0.55	1	135/93	
		28% fat, P/S 0.49	6		+10
		37% fat, P/S 1.42	6		+1/-1
		38% fat, P/S 0.33	6		+2/+1
Sacks et al. (1987b)	17	Double-blind, crossover			
		47% fat, P/S 0.4		117/79	
		23 g linoleic acid	4		-3/-4
		23 g oleic acid	4		+2/+2
Brussard et al. (1981)	60	Parallel			
		30% fat, 11% PUFA	5	124/70	-5/-3
		22% fat, 3% PUFA	5	127/68	-5/-4
		40% fat, 19% PUFA	5	121/68	-5/-3
		39% fat, 3% PUFA	5	118/68	-3/-3
Margetts et al. (1984)		Double-blind, parallel, fixed order			
	16	P/S 0.31-0.38	2+6+6	124/74	-4/-2
	20	P/S 0.48	2	126/78	
		P/S 1.04	6		-3/-4
		P/S 0.33	6		-1/-1
	18	P/S 0.36	2	120/73	-1/-1
		P/S 0.27	6		+1/0
		P/S 1.06	6		

Iacono et al. (1981)	59	Fixed order 39% fat, P/S 0.15 24% fat, P/S 1.2 36% fat, P/S 0.16	1 6 6	125/77	$-8^a/-3^a$ $-8^a/+6^a$
Puska et al. (1983)	38	Parallel, fixed order 37% fat, P/S 0.24	6+4	138/89	0/-1
	35	23% fat, P/S 0.98 37% fat, P/S 0.29	6 4	138/89	$-8^a/-8^a$ $+7^a/+4^a$
Puska et al. (1985)	41	Parallel, fixed order 38% fat, P/S 0.2 23% fat, P/S 0.9 37% fat, P/S 0.2	2 12 2	128/88	$-4^a/-5^a$ $+5^a/+2^a$
	43	38% fat, P/S 0.2 24% fat, P/S 0.4 37% fat, P/S 0.2	2 12 2	126/86	$-3^a/-4^a$ $+3^a/+3^a$

HT = hypertensive   NT = normotensive   P = polyunsaturated   S = saturated

PUFA = polyunsaturated fatty acids

[a] P<0.05 compared to previous diet period   [b] P<0.05 for the difference between the changes in group means   [c] P<0.05 compared to placebo

There are several possible explanations for these inconsistent results. First, it is possible that open dietary interventions have a placebo effect. In the open study by Puska et al. (1983) there was also a salt reduction group (see Table II), which presumably got the same placebo as did the fat intervention group. There was no significant reduction in blood pressure in the low-salt groups. Secondly, since the low fat diet used in the North Karelian studies caused changes in the intake of other nutrients besides fats, it can be argued that the decrease in blood pressure could be due to other dietary changes. However, in the analyses of the pooled data of three studies (Iacono et al., 1983; Puska et al., 1983; Puska et al., 1985), changes in polyunsaturated and saturated fat intake were the strongest predictors of blood pressure change except for initial blood pressure. Changes in intake of total fat, sodium, potassium, magnesium, calcium, carbohydrate or protein had no significant association with blood pressure change (Nissinen et al., 1987). Thirdly, it is possible that the effect of fat on blood pressure is dependent on the P/S ratio of the initial diet. In all the studies giving positive results initial P/S ratios were below 0.3, whereas in the negative studies P/S ratios ranged from 0.3 to 0.55.

The blood pressure-lowering effect of marine oils containing n-3 polyunsaturated fatty acids has been studied by several groups. Cod liver oil supplementation (Lorenz et al., 1983; Mortensen et al., 1983) as well as a mackerel diet (Singer et al., 1983; Singer et al., 1985) have been reported to reduce systolic blood pressure. In the latter studies, the minimum amount of eicosapentanoic acid required for blood pressure reduction was about 2 g/d, which is equivalent to more than 200 g of mackerel daily.

8.3   Mechanism of Blood Pressure Changes

There is evidence that dietary fat influences renal excretory function in both man and animals. An increase in urine volume in linoleic acid deficient rats has been observed in several studies (Rosenthal et al., 1974; Soma et al., 1985; Hansen, 1981). Effects on electrolyte excretion are more controversial. In some studies a reduction in capacity to excrete sodium and potassium during linoleic acid deprivation was observed in salt-loaded rats (Rosenthal et al., 1974), while in others it was not (Smith-Barbaro and Pucak, 1983; Hansen, 1981). Information on humans is limited. A slight increase in sodium and creatinine excretion was seen during consumption of a high linoleic acid diet, but not in water or potassium excretion (Adam and Wolfram, 1984). In contrast, a significant increase in urine volume, but not in sodium excretion, was observed by Puska et al. (1985) in middle-aged males and females fed a low-fat diet with a high P/S ratio.

The most obvious biochemical explanation for the blood pressure changes associated with dietary n-6 and n-3 polyunsaturated fatty acids is an altered metabolism of prostaglandins. Ten Hoor and van de Graaf (1978) reported that the blood pressure-lowering effect of linoleic acid in salt-loaded rats can be blocked by acetylsalicylic acid, an inhibitor of prostaglandin synthesis. Furthermore, high linoleic acid diets raise the tissue level of PGE_2 and 6-keto PGF_1 (the major metabolite of PGI_2) and the urinary excretion of PGE_2 (cf. Pietinen and Huttunen, 1987). At least theoretically, these changes would favor blood pressure reduction, as the vasodilatory compounds PGI_2 and PGE_2 are considered to have an antihypertensive rather than a prohypertensive effect.

Other mechanisms also may be involved in the blood pressure reductions observed in subjects consuming diets rich in polyunsaturated fats. Dietary fats have important effects on the fatty acid composition of cell membranes. These effects may result in alterations in ion permeability, membrane excitability, hormonal response and/or effectiveness of renal and neural pressor mechanisms. Thus, in a study on normotensive volunteers it was recently shown that mean total leucocyte sodium efflux rose significantly during supplementation with linoleic acid as opposed to placebo (Heagerty et al., 1986). Moreover, all components of supine and standing blood pressure fell, though only the fall in supine systolic pressure was significant.

8.4   Comments

Population surveys have provided some support for the hypothesis that dietary fat influences blood pressure regulation in man. The evidence from clinical trials is controversial. A significant reduction in systolic and diastolic blood pressure has been observed in most studies during consumption of low-fat, high P/S diets. Whether these changes are due to an increase in the intake of n-6 polyunsaturated fats, to a simultaneous decrease in saturated fats, or to concomitant changes in the intake of other dietary components, remains to be determined. Increasing only the P/S ratio from an initial level above 0.3 does not reduce blood pressure, suggesting that linoleic acid has an effect on blood pressure only near the deficiency range. This suggestion is supported by animal data.

The effect of n-3 polyunsaturated fats on blood pressure is controversial. Studies on animals have produced both reductions and elevations in blood pressure on high n-3 polyunsaturated fat diets. Results in humans are also inconsistent, but it appears that very high intakes of marine oils may have a blood pressure-lowering effect in man.

9. Vegetarian Diet

Vegetarians generally have lower blood pressures than comparable populations of nonvegetarians (Rouse and Beilin, 1984). Their lower blood pressures, which appear to be determined substantially by their diet, have been attributed to the combined effects of specific nutrients and a lower prevalence of obesity. The foods eaten by vegetarians and by typical omnivora result in substantial differences in nutrient intake. A vegetarian diet has a lower total fat content, a higher P/S ratio, and a higher fiber and vegetable protein content than a mixed diet.

Two controlled trials suggest that vegetarian diets lower blood pressure. An ovo-lactovegetarian diet lowered the blood pressure of both normotensive (Rouse et al., 1983) and mildly hypertensive subjects within 6 weeks (Margetts et al., 1986). Factor analysis suggested that changes in polyunsaturated fats (increase in P/S ratio from 0.29 to 0.68), fiber and protein were most likely to have mediated the observed changes in blood pressure (Rouse et al., 1986). In contrast to these findings, Sacks et al. (1984) found no change in blood pressure after three months of feeding an ovo-lactovegetarian diet. A higher baseline P/S ratio (0.52) and a very low initial blood pressure (116/74 mm Hg) might explain the lack of changes in the participants.

The community benefits of a non-pharmacological reduction of mean blood pressure of the size reported here (5-6 mm Hg systolic) would be substantial. However, more acceptable means than change to a vegetarian diet are needed to make any long-term impact on the prevalence of hypertension. It is clearly important to identify the nutrients responsible for the beneficial effects of a vegetarian diet. None of the characteristics of the vegetarian diet alone—a higher P/S ratio, higher fiber, potassium, magnesium or calcium content—has a convincingly blood pressure-lowering effect. Some preliminary findings suggest that increased fiber intake may reduce blood pressure independently of other dietary factors (Uusitupa et al., 1984; Schlamowitz et al., 1987). There are few data available on the effect of the type of protein or meat in regulating blood pressure, and the effect of a reduction in meat intake separately from other major food intakes has not been studied.

10. Diet as an Alternative to Drug Treatment

Trials on the effects of drug therapy in mild hypertension have generally shown a beneficial effect on risk of stroke, whereas no protective effect against the manifestations of coronary heart disease has been evident (cf. MacMahon *et al.*, 1986). One of the possible explanations for this finding is that certain metabolic effects of antihypertensive drugs may increase the risk of coronary heart disease, thus counteracting the beneficial effects of blood pressure reduction. In mild hypertension, weight reduction alone or in conjunction with other dietary measures such as reduction in salt intake or in alcohol consumption, or increased carbohydrate intake, has caused a reduction of blood pressure. These responses are comparable to the effects of drug therapy, but with the additional benefit of favorable changes in serum cholesterol or triglyceride concentrations or in glucose tolerance (Dodson *et al.*, 1984; MacMahon *et al.*, 1985; Stamler *et al.*, 1987).

A considerable proportion of hypertensive patients who are well managed with drug therapy can be taken off drugs during dietary therapy. Langford *et al.* (1985) found that 72% of obese and 78% of nonobese patients with mild hypertension could be satisfactorily managed with weight reduction and sodium restriction, respectively, during a follow-up period of 56 weeks. Stamler *et al.* (1987) studied the combined effects of weight reduction (in the obese), sodium reduction and reduction in alcohol intake. They found that, among patients whose drug therapy was withdrawn, 39% achieved adequate control of blood pressure for 4 years with diet therapy, compared with only 5% in subjects without diet therapy. In patients able to continue without medication, there was an overall beneficial effect on serum levels of cholesterol, triglycerides, fasting glucose and uric acid, especially among those losing at least 5 lb (2.25 kg). Also, serum potassium levels rose when patients were withdrawn from medication. In contrast, body weight and serum cholesterol levels rose slightly in the control group receiving neither drugs nor diet therapy.

The usual argument against dietary treatment is poor compliance. On average, dropout rates in self-help and other weight loss programs range from 50% to 70% within 1-2 years and only about 50% of patients can be expected to comply with diets prescribed for cardiovascular disease (Dunbar, 1986). However, the results of the Hypertension Control Program (Stamler *et al.*, 1987) are quite promising. The goal for weight loss (at least 4-5 kg or 10 lb) was attained by 30% of subjects and 59% exhibited at least some loss. The goal for sodium intake (80 mmol per day) was reached by 31% of subjects, the average reduction being 36% (from 166 mmol to 106 mmol). Alcohol intake was reduced to about half and the number of heavy drinkers was also halved. There was also a remarkable reduction in fat intake from 39% to 32% of energy. Overall, a majority of subjects were able to make modest dietary changes which were shown to be clearly beneficial.

Although the Dietary Intervention Study in Hypertension Trial (Langford *et al.*, 1985) dealt with sodium reduction and weight reduction as single interventions in distinct groups, and did not include alcohol reduction as a goal, the similarities in the results of the two trials reinforce the conclusion that diet therapy is an important and useful strategy in the control of hypertension. Since atherosclerotic complications are the major cause of mortality in hypertensive patients, the role of diet in the treatment of hypertensives should not be underestimated.

11. Conclusions

Evidence from numerous studies indicates that it is possible to reduce high blood pressure by various non-pharmacologic means. For prevention of hypertension, control of body weight seems essential, as an increase in body weight during middle age shows a particularly clear association with the development of hypertension. Another preventive measure with a definite effect on blood pressure at the population level is reduction of excessive alcohol intake. As pointed out by Feinstein (1985), in large population studies the influence of age, body mass index and alcohol intake on blood pressure is far greater than that of individual nutrients such as different minerals.

For reduction of increased blood pressure by nutritional means, weight reduction evidently is the most effective method. A decrease by 5-10 mm Hg in both systolic and diastolic blood pressure can be expected in overweight hypertensive individuals who experience the degree of weight reduction which has been generally achieved. Some individuals may get far better results from greater weight reduction. Restriction of alcohol consumption has resulted in decrements of the order of 5 mm Hg in both systolic and diastolic blood pressure. A feasible restriction of sodium intake will result in a mean reduction of 2-3 mm Hg even in individuals of normal weight. The response to sodium restriction is variable at the individual level and better overall effects may be seen in salt-sensitive subjects.

These three approaches to nonpharmacological control of high blood pressure were recommended in the recent report of the Joint National Committee on Detection, Evaluation, and Treatment of High Blood Pressure in the U.S.A. (NIH, 1986). Evidence on the efficacy of other measures, such as supplementation with potassium, calcium, magnesium and dietary fibre or a change in fat intake, is still insufficient to justify a general statement or recommendation. However, as shown by the effects of the vegetarian diet, a combination of dietary measures has proven more effective than changes in the consumption of single nutrients. It is apparent that a majority of Western populations would never comply with a vegetarian diet. However, changes of dietary habits in favor of foods of plant origin may reduce blood pressure by combining small beneficial effects of several dietary factors.

The advantages of a nonpharmacological approach are evident in the treatment of mild hypertension. For the ultimate goal of antihypertensive treatment (the prevention of cardiovascular complications), it would seem prudent to advocate dietary therapy if only as an adjunct to antihypertensive drug therapy. Weight reduction is effective even in drug-treated hypertensive subjects. Alcohol consumption increases blood pressure and reduces compliance with antihypertensive therapy in drug-treated hypertensive patients. The benefits of diet therapy include favorable metabolic effects and a reduced need for antihypertensive drug therapy.

References

Adam, O., and Wolfram, G., 1984, Effect of different linoleic acid intakes on prostaglandin biosynthesis and kidney function in man, *Am. J. Clin. Nutr.* 40:763.

Addison, W. L. T., 1928, The use of sodium chloride, potassium chloride, sodium bromide and potassium bromide in cases of arterial hypertension which are amenable to potassium chloride, *Can. Med. Assoc. J.* 181:281.

Aro, A., 1987, Dietary calcium and hypertension: population studies, *Eur. Heart J.* 8 (Suppl. B):31.

Barden, A. E., Vandongen, R., Beilin, L. J., Margetts, B., and Rogers, P., 1986, Potassium supplementation does not lower blood pressure in normotensive women, *J. Hypertension* 4:339.

Beard, T. C., Cooke, H. M., and Barge, R., 1982, Randomised controlled trial of a no-added-sodium diet for mild hypertension, *Lancet* 2:455.

Beevers, D. G., Zezulka, A. V., Potter, J. F., Bannan, L. T., Maheswaran, R., and Gill, J. S., 1987, The clinical relevance of alcohol in the blood pressure clinic, *Eur. Heart J.* 8 (Suppl. B):27.

Belizan, J. M., Villar, J., Pineda, O., Gonzales, A. E., Sainz, E., Garrera, G., and Sibrian, R., 1983, Reduction of blood pressure with calcium supplementation in young adults, *JAMA* 249:1161.

van Beresteyn, E. C. H., Schaafsma, G. and de Waard, H., 1986, Oral calcium and blood pressure: a controlled intervention trial, *Am. J. Clin. Nutr.* 44:883.

Berry, E. M., and Hirsch, J., 1986, Does dietary linolenic acid influence blood pressure? *Am. J. Clin. Nutr.* 44:336.

Bloomfield, R. L., Young, L., Zurek, G., Felts, J. H., and Straw, M. K., 1986, Effects of oral calcium carbonate on blood pressure in subjects with mildly elevated arterial pressure, *J. Hypertension* 4 (Suppl. 5):S351.

Brussaard, J. H., van Raaij, J. M. A., StasseG.-Wolthuis, M., Katan, M. B., and Hautvast, J. G., 1981, Blood pressure and diet in normotensive volunteers: absence of an effect of dietary fiber, protein, or fat. *Am. J. Clin. Nutr.* 34:2023.

Bulpitt, D. J., Daymond, M., Bulpitt, P. F., Ferrier, G., Harrison, R., Lewis, P. J., Dollery, P. J., and Dollery, C. T., 1984, Is low salt dietary advice a useful therapy in hypertensive patients with poorly controlled blood pressure? *Ann. Clin. Res.* 16 (Suppl. 43):143.

Bulpitt, D. J., Ferrier, P. J., Lewis, M., Bulpitt, P. F., and Dollery, C. T., 1985, Potassium supplementation fails to lower blood pressure in hypertensive patients receiving a potassium losing diuretic, *Ann. Clin. Res.* **17:**126.

Burstyn, P., Hornall, D., and Watchorn, C., 1980, Sodium and potassium intake and blood pressure, *Br. Med. J.* **2:**537.

Cappuccio, F. P., Markandu, N. D., Beynon, G. W., Shore, A. C., Sampson, B., and MacGregor, G. A., 1985, Lack of effect of oral magnesium on high blood pressure: a double blind study, *Br. Med. J.* **291:**235.

Cappuccio, F. P., Markandu, N. D., Singer, D. R. J., Smith, S. J., Shore, A. C., and MacGregor, G. A., 1987, Does oral calcium supplementation lower high blood pressure? A double blind study, *J. Hypertension* **5:**67.

Carney, S., Morgan, T., Wilson, M., Matthews, G., and Roberts, R., 1975, Sodium restriction and thiazide diuretics in the treatment of hypertension, *Med. J. Aust.* **1:**1803.

Chalmers, J., Morgan, T., Doyle, A., Dickson, B., Hopper, J., Matthews, J., Matthews, G., Moulds, R., Myers, J., Nowson, C., Scoggins, B., and Stebbing, M., 1986, Australian National Health and Medical Research Council dietary study in mild hypertension, *J. Hypertension* **4** (Suppl. 6):S629.

Chapman, C. B., and Gibbons, T. B., 1949, The diet and hypertension: a review, *Medicine* **29:**29.

Chiang, B. N., Perlman, L. V., and Epstein, F. H., 1969, Overweight and hypertension: a review, *Circulation* **34:**403.

Cohen, L., Laår, A., and Kitzes, R., 1984, Reversible retinal vasospasms in magnesium-treated hypertension despite no significant change in blood pressure, *Magnesium, Exp. Clin. Res.* **3:**159.

Comberg, H. V., Heyden, S., Hames, C. G., Vergroesen, A. J., and Fleischman, A. I., 1978, Hypotensive effect of dietary prostaglandin precursors in hypertensive man, *Prostaglandins* **15:**193.

Cooke, K. M., Frost, G. W., and Stokes, G. S., 1983, Blood pressure and its relationship to low levels of alcohol consumption, *Clin. Exp. Pharmacol. Physiol.* **10:**229.

Cooke, K. M., Frost, G. W., Thornell, I. R., and Stokes, G. S., 1982, Alcohol consumption and blood pressure: survey of the relationship at a health screening clinic, *Med. J. Aust.*, **1:**65.

Criqui, M. H., 1987, Alcohol and hypertension: New insights from population studies, *Eur. Heart J.* **8** (Suppl. B):19.

Criqui, M. H., Wallace, R. B., Mishkel, M., Barrett-Connor, E., and Heiss, G., 1981, Alcohol consumption and blood pressure: the Lipid Research Clinics Prevalence Study, *Hypertension* **3:**557.

Cruickshank, J. K., Jackson, S. H. D., Beevers, D. G., Bannan, L. T., Beevers, M., and Stewart, V. L., 1985, Similarity of blood pressure in blacks, whites and Asians in England: the Birmingham Factory Study, *J. Hypertension* **3:**365.

Dahl, L. K., Silver, L., Christie, R. W., 1958, The role of salt in the fall of blood pressure accompanying reduction in obesity, *N. Engl. J. Med.*, **258:**1186.

Dodson, P. M., Pacy, P. J., Bal, P., Kubicki, A. J., Fletcher, R. F., and Taylor, K. G., 1984, A controlled trial of a high-fibre, low fat and low sodium diet for mild hypertension in type 2 (non-insulin-dependent) diabetic patients. *Diabetologia* **27:**522.

Dornfeld, L. P., Maxwell, M. H., Waks, A. U., Schroth, P., and Tuck, M. L., 1985, Obesity and hypertension: long-term effects of weight reduction on blood pressure, *Int. J. Obesity* **9:**381.

Dunbar, J., 1986, Practical aspects of dietary management of hypertension: compliance, *Can. J. Physiol. Pharmacol.* **64**:831.

Dyckner, T., and Wester, P. O., 1983, Effect of magnesium on blood pressure, *Br. Med. J. (Clin. Res.)* **286**:1847.

Eliahou, H. E., Iaina, A., Gaon, T., Shochat, J., and Modan, M., 1981, Body weight reduction necessary to attain normotension in the overweight hypertensive patient, *Int. J. Obesity* **5** (Suppl. 1):157.

Elliott, P., Fehily, A. M., Sweetnam, P. M., and Yarnell, J. W. G., 1987, Diet, alcohol, body mass, and social factors in relation to blood pressure: the Caerphilly Heart Study, *J. Epidemiol. Community Health* **41**:37.

Erwteman, T. M., Nagelkarke, N., Lubsen, J., Koster, M., and Dunning, A. J., 1984, β-blockade, diuretics, and salt restriction for the management of mild hypertension: a randomised double blind trial, *Br. Med. J.* **289**:406.

Fagerberg, B., Andersson, O. K., Isaksson, B., and Björntorp, P., 1984, Blood pressure control during weight reduction in obese hypertensive men: separate effects of sodium and energy restriction, *Br. Med. J.* **288**:11.

Feinstein, A. R., 1985, Tempest in a p-pot? *Hypertension* **7**:313.

Friedman, G. D., Klatsky, A. L., and Siegelaub, M. S., 1983, Alcohol intake and hypertension, *Ann. Intern. Med.* **98**:846.

Froment, A., Milon, H., and Gravier, C., 1979, Relation entre consonimation sondee et hypertension arterielle: contribution de l'epidemiologie geographique, *Rev. Epidemiol. Santé Publique* **27**:437.

Gleiberman, L., 1973, Blood pressure and dietary salt in human populations, *Ecol. Food Nutr.* **2**:143.

Grim, C. E., Luft, F. C., Miller, J. Z., Meneely, G. R., Battarbee, H. D., Hames, C. G., and Dahl, H. K., 1980, Racial differences in blood pressure in Evans County, Georgia: relationship to sodium and potassium intake and plasma renin activity, *J. Chron. Dis.* **33**:87.

Grobbee, D. E., and Hofman, A., 1986, Effect of calcium supplementation on diastolic blood pressure in young people with mild hypertension, *Lancet* **2**:703.

Grobbee, D. E., Hofman, A., and Roelandt, J. T., Boomsma, F., Schalekamp, M. A., and Valkenburg, H. A., 1987, Sodium restriction and potassium supplementation in young people with mildly elevated blood pressure, *J. Hypertension* **5**:115.

Grollman, A., 1930, The action of alcohol, caffeine and tobacco on cardiac output (and its related functions) of normal man, *J. Pharmacol.* **39**:313.

Gruchow, H. W., Sobocinski, K. A., and Barboriak, J. J., 1985, Alcohol, nutrient intake, and hypertension in US adults, *JAMA* **253**:1567.

Guyton, A. C., Granger, H. J., and Coleman, T. G., 1971, Auto-regulation of the total systemic circulation and its relation to control of cardiac output and arterial pressure, *Circ. Res.* **28**:1.

Hansen, H. S., 1981, Essential fatty acid supplemented diet increases renal excretion of prostaglandin E and water in essential fatty acid deficient rats, *Lipids* **16**:849.

Harlan, W. R., and Harlan, L. C., 1986, An epidemiological perspective on dietary electrolytes and hypertension, *J. Hypertension* **4** (Suppl. 5):S334.

Harlan, W. R., Hull, A. L., Schmouder, R. L., Landis, J. R., Thompson, F. E., and Larkin, F. A., 1984, Blood pressure and nutrition in adults: the National Health and Nutrition Examination Survey, *Am. J. Epidemiol.* **120**:17.

Haynes, R. B., Harper, A. C., Costely, S. R., Johnston, M., Logan, A. G., Flanagan, P. T., and Sackett, D. L., 1984, Failure of weight reduction to reduce mildly elevated blood pressure: a randomized trial, *J. Hypertension* **2**:535.

Heagerty, A. M., Ollerenshaw, J. D., Robertson, D. I., Bing, R. F., and Swales, J. D., 1986, Influence of dietary linoleic acid on leucocyte sodium transport and blood pressure, *Br. Med. J.* **293**:295.

Henderson, D. G., Schierup, J., and Schodt, T., 1986, Effect of magnesium supplementation on blood pressure and electrolyte concentrations in hypertensive patients receiving long term diuretic treatment, *Br. Med. J.* **293**:664.

Hofman, A., Hazebroek, A., and Valkenburg, H. A., 1983, A randomized trial of sodium intake and blood pressure in newborn infants, *JAMA* **250**:370.

Holly, J. M. P., Goodwin, F. J., Evans, S. J. W., Van den Burg, M. J., and Ledingham, J. M., 1981, Re-analysis of data in two Lancet papers on the effect of dietary sodium and potassium on blood pressure, *Lancet* **2**:1384.

ten Hoor, F., and van de Graaf, H. M., 1978, The influence of a linoleic acid-rich diet and of salicylic acid on NaCl induced hypertension, Na- and H_2O-balance and urinary prostaglandin excretion in rats, *Acta Biol. Med. Germ.* **37**:875.

Hovell, M. F., 1982, The experimental evidence for weight-loss treatment of essential hypertension: a critical review, *Am. J. Publ. Health* **72**:359.

Hsu, P-H., Mathewson, F. A. L., and Rabkin, S. W., 1977, Blood pressure and body mass index patterns--a longitudinal study, *J. Chron. Dis.* **30**:93.

Iacono, J. M., Judd, J. T., Marshall, M. W., Canary, J. J., Dougherty, R. M., Mackin, J. F., and Weinland, B. T., 1981, The role of dietary essential fatty acids and prostaglandins in reducing blood pressure, *Progr. Lipid Res.* **20**:349.

Iacono, J. M., Marshall, M. W., Dougherty, R. M., Wheeler, M. A., Mackin, J. F., and Canary, J. J., 1975, Reduction in blood pressure associated with high poly-unsaturated fat diets that reduce blood cholesterol in man, *Prev. Med.* **4**:426.

Iacono, J. M., Puska, P., Dougherty, R. M., Pietinen, P., Variainen, E., Leino, U., Mutanen, M.,, and Moisio, S., 1983, Effect of dietary fat on blood pressure in a rural Finnish population, *Am. J. Clin. Nutr.* **38**:860.

Iimura, O., Kijima, T., Kikuchi, K., Miama, A., Ando, T., Nakao, T., and Takigami, Y., 1981, Studies on the hypotensive effect of high potassium intake in patients with essential hypertension, *Clin. Sci.* **61** (Suppl.):77.

Imai, Y., Sato, K., Abe, K., Sasaki, S., Nihei, M., Yoshinaga, K., Sekino, H., 1986, Effect of weight loss on blood pressure and drug consumption in normal weight patients, *Hypertension* **8**:223-228.

INTERSALT Co-operative Research Group, 1988, INTERSALT: An international study of electrolyte excretion and blood pressure: Results of 24-hour urinary sodium and potassium, *Br. Med. J.* **297**:319.

Joffres, M. R., Reed, D. M., and Yano, K., 1987, Relationship of magnesium intake and other dietary factors to blood pressure: the Honolulu heart study, *Am. J. Clin. Nutr.* **45**:469.

Johnson, N. E., Smith, E. L., and Freudenheim, J. L., 1985, Effects on blood pressure of calcium supplementation of women, *Am. J. Clin. Nutr.* **42**:12.

Judd, J. T., Marshall, M. W., and Canary, J., 1981, Effects of diets varying in fat and P/S-ratio on blood pressure and blood lipids in adult men, *Progr. Lipid Res.* **20**:571.

Kannel, W., Brand, N., Skinner, J., Dawber, T., and McNamara, P., 1976, Relation of adiposity to blood pressure and development of hypertension: the Framingham Study, *Ann. Intern. Med.* **68**:48.

Kaplan, N. M., 1985, Non-drug treatment of hypertension, *Ann. Intern. Med.* **103**:359.

Kaplan, N. M., Carnegie, A., Raskin, P., Heller, J. A., and Simmons, M., 1985, Potassium supplementation in hypertensive patients with diuretic-induced hypokalemia, *N. Engl. J. Med.* **312**:746.

Kesteloot, H., 1984, Urinary cations and blood pressure-population studies, *Ann. Clin. Res.* **16** (Suppl. 43):72.

Khaw, K.-T., and Barret-Connor, E., 1987, Dietary potassium and stroke-associated mortality, *N. Engl. J. Med.* **316**:235.

Kihara, M., Fujikawa, J., Ohtaka, M., Mano, M., Nara, Y., Horie, R., Tsunematsu, T., Note, S., Fukase, M., and Yamori, Y., 1984, Interrelationships between blood pressure, sodium, potassium, serum cholesterol, and protein intake in Japanese, *Hypertension* **6**:736.

Klatsky, A. L., Friedman, G. C., and Armstrong, M. A., 1986, The relationship between alcoholic beverage use and other traits to blood pressure: a new Kaiser Permanente study, *Circulation* **73**:628.

Klatsky, A. L., Friedman, G. D., Siegelaub, A. B., and Gerard, M. J., 1977, Alcohol consumption and blood pressure: Kaiser-Permanente multiphasic health examination data, *N. Engl. J. Med.* **296**:1194.

Kok, F. J., Vandenbroucke, J. P., van der Heide-Wessel, C., and van der Heide, R. M., 1986, Dietary sodium, calcium, and potassium, and blood pressure, *Am. J. Epidemiol.* **123**:1043.

Kromhout, D., Bosschieter, E. B., and de Lezenne Coulander, C., 1985, Potassium, calcium, alcohol intake and blood pressure: the Zutphen Study, *Am. J. Clin. Nutr.* **41**:1299.

Kupari, M., 1983, Acute cardiovascular effects of alcohol, *Br. Heart J.* **49**:174.

Kupari, M., Heikkilä, J., Tolppanen, E.-M., Nieminen, M. S., and Ylikahri, R., 1983, Acute effects of alcohol, beta blockade, and their combination on left ventricular function and haemodynamics in normal man, *Eur. Heart J.* **4**:463.

Lang, T., Degoulet, P., Aime, F., Devries, C., Jacquinet-Salord, M.-C., and Fouriaud, C., 1987, Relationship between alcohol consumption and hypertension prevalence and control in a French population, *J. Chron. Dis.* **40**:713.

Langford, H. G. 1986, The passive role of calcium in hypertension: a position statement as of August 20, 1985, *Can. J. Physiol. Pharmacol.* **64**:808.

Langford, H. G., Blaufox, M. D., Oberman, A., Hawkins, C. M., Curb, J. B., Cutter, G. R., Wassertheil-Smoller, S., Pressel, S., Babcock, C., Abernethy, J. D., Hotchkiss, J., and Tyler, M., 1985, Dietary therapy slows the return of hypertension after stopping prolonged medication, *JAMA* **253**:657.

Laragh, J. H., and Pecker, M. S., 1983, Dietary sodium and essential hypertension: some myths, hopes, and truths. *Ann. Intern. Med.* **98**:735.

Levi, R. L., White, P. D., and Shoud, W. D., 1946, Overweight: a prognostic significance in relation to hypertension and cardiovascular renal disease, *JAMA* **131**:951.

Ljunghall, S., Hvarfner, A., and Lind, L., 1987, Clinical studies of calcium metabolism in essential hypertension, *Eur. Heart J.* **8** (Suppl. B):37.

Lorenz, R., Spengler, U., Fischer, S., Duhm, J., and Weber, P. C., 1983, Platelet function, thromboxane formation and blood pressure control during supplementation of the Western diet with cod liver oil, *Circulation* **67**:504.

Lyle, R. M., Melby, C. L., Hyber, G. C., Edmondson, J. W., Miller, J. Z., and Weinberger, M. H., 1987, Blood pressure and metabolic effects of calcium supplementation in normotensive white and black men, *JAMA* **257**:1772.

MacGregor, G. A., 1985, Sodium is more important than calcium in essential hypertension, *Hypertension* **7**:628.

MacGregor, G. A., Markandu, N. D., Best, F. E., Elder, D. M., Cam, J. M., Sagnella, G. A., and Squires, M., 1982, Double-blind randomised crossover trial of moderate sodium restriction in essential hypertension, *Lancet* **1**:352.

MacMahon, S. W., Blacket, R. B., Macdonald, G. J., and Hall, W., 1984, Obesity, alcohol consumption and blood pressure in Australian men and women: the National Heart Foundation of Australia risk factor prevalence study, *J. Hypertension* 2:85.

MacMahon, S. W., Macdonald, G. J., Bernstein, L., Andrews, G., and Blacket, R. B., 1985, Comparison of weight reduction with metoprolol in treatment of hypertension in young overweight patients, *Lancet* 1:1233.

MacMahon, S. W., Cutler, J. A., Furberg, C. D., and Payne, G. H., 1986, The effects of drug treatment for hypertension on morbidity and mortality from cardiovascular disease: a review of randomized controlled trials, *Prog. Cardiovasc. Dis.* 24 (Suppl. 1):99.

Malhotra, H., Mehta, S. R., Mathur, D., Khandelwal, P. D., 1985, Pressor effects of alcohol in normotensive and hypertensive subjects, *Lancet* 2:584.

Margetts, B. M., Beilin, L. J., Armstrong, B. K., Vandongen, R., and Croft, K. D., 1984, Dietary fat intake and blood pressure: a double blind controlled trial of changing polyunsaturated to saturated fat ratio, *J. Hypertension* 2 (Suppl. 3):201.

Margetts, B. M., Beilin, L., Vandongen, R., and Armstrong, B. K., 1986, Vegetarian diet in mild hypertension: a randomized controlled trial, *Br. Med. J.* 293:1468.

Mathews, J. D., 1976. Alcohol use, hypertension, and coronary heart disease, *Clin. Sci.* 51:661S.

Matlou, S. M., Isles, C. G., Higgs, A., Milne, F. J., Murray, G. D., Schultz, E., and Starke, E. F., 1986, Potassium supplementation in blacks with mild to moderate essential hypertension, *J. Hypertension* 4:61.

Maxwell, M. H., Kushiro, T., Dornfeld, L. P., Tuck, M. L., and Waks, A. U., 1984, BP changes in obese hypertensive subjects during rapid weight loss: Comparison of restricted v. unchanged salt intake, *Arch. Intern. Med.* 144:1581.

M'Buyamba-Kabangu, J. R., Fogard, R., Lijnen, P., Mbuy wa Mbuy, R., Staessen, J., and Amery, A., 1986, Blood pressure and urinary cations in urban Bantu of Zaire, *Am. J. Epidemiol.* 124:957.

McCarron, D. A., Morris, C. D., Henry, H. J., Stanton, J. L., 1984, Blood pressure and the nutrient intake in the United States, *Science* 224:1392.

McCarron, D. A., and Morris, C. D., 1985, Blood pressure response to oral calcium in persons with mild to moderate hypertension: a randomized, double-blind, placebo-controlled, crossover trial, *Ann. Intern. Med.* 103:825.

McCarron, D. A., Pingree, P. A., Rubin, R. J., Gaucher, S. M., Molitch, M., and Kritzik, S., 1980, Enhanced parathyroid function in essential hypertension: a homeostatic response to a urinary calcium leak, *Hypertension* 2:162.

Meneely, G. R., and Dahl, I. K., 1961, Electrolytes in hypertension: the effects of sodium chloride, *Med. Clin. N. Am.* 45:271.

Miettinen, T. A., Naukkarinen, V., Huttunen, J. K., Mattila, S., and Kumlin, T., 1982, Fatty-acid composition of serum lipids predicts myocardial infarction, *Br. Med. J.* 285:993.

Morgan, T., Adam, W., Gillies, A., Wilson, M., Morgan, G., and Carney, S., 1978, Hypertension treated by salt restriction, *Lancet* 1:227.

Mortensen, J. A., Schmidt, E. B., Nielsen, A. H., and Dyerberg, J., 1983, The effect of n-6 and n-3 polyunsaturated fatty acids on hemostasis, blood lipids and blood pressure, *Thromb. Haemost.* 50:543.

Nichaman, M., Shekelle, R., and Paul, O., 1984, Diet, alcohol, and blood pressure in the Western Electric Study, *Am. J. Epidemiol.* 120:469.

NIH, 1986, Nonpharmacological approaches to the control of high blood pressure: Final report of the Subcommittee on Nonpharmacological Therapy of the 1984

Joint National Committee on Detection, Evaluation, and Treatment of High Blood Pressure, *Hypertension* 8:444.

Nissinen, A., Pietinen, P., Tuomilehto, J., Vartiainen, E., Iacono, J., and Puska, P., 1987, Predictors of blood pressure change in a series of controlled dietary intervention studies, *J. Human Hypertension* (in press).

Nowson, C., and Morgan, T., 1986, Effect of calcium carbonate on blood pressure, *J. Hypertension* 4 (Suppl. 5):S673.

Orlando, J., Aronow, W. S., Cassidy, J. , and Prakash, R., 1976, Effect of ethanol on angina pectoris, *Ann. Intern. Med.* 84:652.

Oster, P., Arab, L., Schellenberg, B., Heuck, C. C., Mordasini, R., and Schlierf, G., 1979, Blood pressure and adipose tissue linoleic acid, *Res. Exp. Med. (Berl)* 175:287.

Overlack, A., Muller, H.-M., Kolloch, R., Ollig, A., Moch, B., Kleinmann, R., Kruck, F., and Stumpe, K. O., 1983, Long-term antihypertensive effect of oral potassium in essential hypertension, *J. Hypertension* 1 (Suppl. 2):165.

Owens, C. J., and Brackett, N. C., Jr., 1987, Role of sodium intake in the antihypertensive effect of propranolol, *South Med. J.* 71:43.

Parfrey, P. S., Condon, K., Wright, P., Vandenburg, M. J., Holly, J. M. P., Goodwin, F. J., Evans, S. J. W., and Ledingham, J. M., 1981, Blood pressure and hormonal changes following alteration in dietary sodium and potassium in young men with and without a familial predisposition to hypertension, *Lancet* 1:113.

Parijs, J., Joossens, J. V., Van der Linden, L., Verstreken, G., and Amery, A., 1973, Moderate sodium restriction and diuretics in the treatment of hypertension, *Am. Heart J.* 85:22.

Pietinen, P., and Huttunen, J. K., 1987, Dietary fat and blood pressure—a review, *Eur. Heart J.* 8 (Suppl. B):9.

Postnov, Y. V., and Orlov, S. N., 1984, Cell membrane alteration as a source of primary hypertension, *J. Hypertension* 2:1.

Potter, J. F., and Beevers, D. G., 1984, Pressor effect of alcohol in hypertension, *Lancet* 1:119.

Potter, J. F., Watson, R. D. S., Skan, W., and Beevers, D. G., 1986, The pressor and metabolic effects of alcohol in normotensive subjects, *Hypertension* 8:625.

Puddey, I. B., Beilin, L. B., and Vandongen, R., 1987, Regular alcohol use raises blood pressure in treated hypertensive subjects: a randomised controlled trial, *Lancet* 1:647.

Puddey, I. B., Beilin, L. J., Vandongen, R., Rouse, I. L., and Rogers, P., 1985a, Evidence for a direct effect of alcohol consumption on blood pressure in normotensive men: a randomized controlled trial, *Hypertension* 7:707.

Puddey, I. B., Vandongen, R., Beilin, L. J., Rouse, I. L., 1985b, Alcohol stimulation of renin release in man: its relation to the hemodynamic, electrolyte, and sympatho-adrenal responses to drinking, *J. Clin. Endocrinol. Metab.*, 61:37.

Puska, P., Iacono, J. M., Nissinen, A., Korhonen, H., Vartiainen, E., Pietinen, P., Dougherty, R., Leino, U., Mutanen, M., Moisio, S., and Huttunen, J. K., 1983. Controlled randomized trial of the effect of dietary fat on blood pressure, *Lancet* 1:1.

Puska, P., Iacono, J. M., Nissinen, A., Vartiainen, E., Dougherty, R., Pietinen, P., Leino, U., Uusitalo, U., Kuusi, R., Kostiainen, E., Nikkari, T., Seppälä, E., Vapaatalo, H., and Huttunen, J. K., 1985, Dietary fat and blood pressure: an intervention study on the effects of a low-fat diet with two levels of polyunsaturated fat, *Prev. Med.* 14:573.

Raison, J., Achimastos, A., Bouthier, J., London, G., Safar, M., 1983, Intravascular volume, extracellular fluid volume, and total body water in obese and nonobese hypertensive patients, *Am. J. Cardiol.* **51:**165.

Ram, C. V. S., Garrett, B. N., and Kaplan, N. M., 1981, Moderate sodium restriction and various diuretics in the treatment of hypertension: effects of potassium wastage and blood pressure, *Arch. Intern. Med.* **141:**1015.

Ramsay, L. E., Ramsay, M. H., Hettiarachchi, J., Davies, D. L., Winchester, J., 1978, Weight reduction in a blood pressure clinic, *Br. Med. J.* **2:**224.

Rao, R. H., Rao, U. B., and Srikantia, S. G., 1981, Effect of polyunsaturate-rich vegetable oils on blood pressure in essential hypertension, *Clin. Exp. Hypertension* **3:**27.

Reaven, G. M., and Hoffman, B. B., 1987, A role for insulin in the aetiology and course of hypertension, *Lancet* **2:**435.

Reed, D., McGee, D., Yano, K., and Hankin, J., 1985, Diet, blood pressure, and multicollinearity, *Hypertension* **7:**405.

Reisin, E., and Frolich, E. D., 1982, Effects of weight reduction on arterial pressure, *J. Chron. Dis.* **35:**887.

Reisin, E., Abel, R., Modan, M., Silverberg, D. S., Eliahou, H. E., and Modan, B., 1978, Effect of weight loss without salt restriction on the reduction of blood pressure in overweight hypertensive patients, *N. Engl. J. Med* **298:**1.

Reisin, E., Frohlich, E. D., Messerli, F. H., Dreslinski, G. R., Dunn, F. G., Jones, M. M., and Batson, H. M., Jr., 1983, Cardiovascular changes after weight reduction in obesity hypertension, *Ann. Intern. Med.* **98:**315.

Resnick, L. M., DiFabio, B., Marion, R., James, G. D., and Laragh, J. H., 1986a, dietary calcium modifies the pressor effects of dietary salt intake in essential hypertension, *J. Hypertension* 4 (Suppl. 5):S679.

Resnick, L. M., Muller, F. B., and Laragh, J. H., 1986b, Calcium-regulating hormones in essential hypertension: Relation to plasma renin activity and sodium metabolism, *Ann. Intern. Med.* **105:**649.

Richards, A. M., Nicholls, M. G., Espiner, E. A., Ikram, H., Maslowski, A. H., Hamilton, E. J., and Wells, J. E., 1984, Blood-pressure response to moderate sodium restriction and to potassium supplementation in mild essential hypertension, *Lancet* **1:**757.

Riemersma, R. A., Wood, D. A., Butler, S., Elton, R. A., Oliver, M., Salo, M., Nikkari, T., Vartiainen, E., Puska, P., Gey, F., Rubba, P., Mancini, M., and Fidanza, F., 1986, Linoleic acid content in adipose tissue and coronary heart disease, *Br. Med. J.* **292:**1423.

Rissanen, A., Pietinen, P., Siljamäki-Ojansuu, U., Piirainen, H., and Reissel, P., 1985, Treatment of hypertension in obese patients: efficacy and feasibility of weight and salt reduction programs, *Acta Med. Scand.* **218:**149.

Robinson, B. F., 1984, Altered calcium handling as a cause of primary hypertension, *J. Hypertension* **2:**453.

Rosenthal, J., Simone, P. G., and Silbergleit, A., 1974, Effects of prostaglandin deficiency on natriuresis, diuresis, and blood pressure, *Prostaglandins* **5:**435.

Rouse, I. K., Beilin, L. J., Armstrong, B. K., and Vandongen, R., 1983, Blood pressure-lowering effect of a vegetarian diet: controlled trial in normotensive subjects, *Lancet* **1:**5.

Rouse, I. K., Beilin, L. J., Mahoney, D. P., Margetts, B. M., Armstrong, B. K., Record, S. J., Vandongen, R., and Barden, A., 1986, Nutrient intake, blood pressure, serum and urinary prostaglandins and serum thromboxane B_2 in a controlled trial with a lacto-ovo-vegetarian diet, *J. Hypertension* **4:**241.

Sacks, F. M., Marais, G. E., Handysides, G., Solazar, J., Miller, L., Foster, J. M., Rosner, B., and Kass, E., 1984, Lack of an effect of dietary saturated fat and cholesterol on blood pressure in normotensives, *Hypertension* 6:193.

Sacks, F. M., Rouse, J. L., Stampfer, M. J., Bishop, L. M., Lenherr, C. F., and Walther, R. J., 1987a, Effect of dietary fats and carbohydrate on blood pressure of mildly hypertensive patients, *Hypertension* 10:452.

Sacks, F. M., Stampher, M. J., Munoz, A., McManus, K., Canessa, M., and Kass, E., 1987b, Effect of linoleic and oleic acids on blood pressure, blood viscosity, and erythrocyte cation transport, *J. Am. Coll. Nutr.* 6:179,

Salonen, J. T., Tuomilehto, J., and Tanskanen, A., 1983, Relation of blood pressure to reported intake of salt, saturated fats and alcohol in healthy nmiddle-aged population, *J. Epidemiol. Comm. Health* 37:32.

Schlamowitz, P., Halberg, T., Warnoe, O., Wilsrtup, F., and Ryttig, K, 1987, Treatment of mild to moderate hypertension with dietary fibre, *Lancet* 2:622.

Sempos, C., Cooper, R., Kovar, M. G., Johnson, C., Drizd, T., and Yetley, E., 1986, Dietary calcium and blood pressure in National Health and Nutrition Examination Surveys I and II, *Hypertension* 8:1067.

Siani, A., Strazzullo, P., Russo, L., Guglielmi, S., Iacoviello, L., Ferrara, L. A., and Mancini, M., 1987, Controlled trial of long term oral potassium supplements in patients with mild hypertension, *Br. Med. J.* 294:1453.

Silman, A. J., Locke, C., Mitchell, P., and Humpherson, P., 1983, Evaluation of the effectiveness of a low sodium diet in the treatment of mild to moderate hypertension, *Lancet* 1:1179.

Simpson, F. O., 1985, Blood pressure and sodium intake, in: *The Epidemiology of Hypertension* (C. J. Bulpitt, ed.), pp. 175-190, Handbook of Hypertension, vol. 4, Elsevier, Amsterdam.

Singer, P., Jaeger, W., Wirth, M., Voigt, S., Nauman, E., Zimontkowski, S., Hajdu, I., and Goedicke, W., 1983, Lipid and blood pressure-lowering effect of mackerel diet in man, *Atherosclerosis* 49:99.

Singer, P., Wirth, M., Voigt, S., Richter-Heinrick, E., Gödicke, W., Berger, I., Nauman, E., Listing, J., Hartrodt, W., and Taube, C., 1985, Blood pressure- and lipid-lowering effect of mackerel and herring diet in patients with mild hypertension, *Atherosclerosis* 56:223.

Skrabal, F., Auböck, J., and Hörtnage, H., 1981, Low sodium/high potassium diet for prevention of hypertension: probable mechanisms of action, *Lancet* 2:895.

Skrabal, F., Gasser, R. W., Finkenstedt, G., Rhomberg, H. P., and Lochs, A., 1984, Low-sodium diet versus low-sodium/high-potassium diet for treatment of hypertension, *Klin. Wochenschr.* 62:124.

Smith, S. J., Markandu, N. D., Sagnells, G. A. amd MacGregor, G. A., 1985, Moderate potassium chloride supplementation in essential hypertension: is it additive to moderate sodium restriction? *Br. Med. J.* 290:110.

Smith-Barbaro, P. A., and Pucak, G. J., 1983, Dietary fat and blood pressure, *Ann. Int. Med.* 98:828.

Soma, M., Manhu, M. S., Jenkins, D. K., Huang, Y. S., and Horrobin, D. F., 1985, Effects of dietary saturated, n-6 and n-3 polyunsaturated fats on blood pressure and prostaglandin outflow from perfused mesenteric vascular beds in rats, *Prostaglandins* 30:969.

Sowers, J. R., Nyby, M., Stern, N., Beck, F., Baron, S., Catania, R. amd Vlachis, N., 1982, Blood pressure and hormone changes associated with weight reduction in the obese, *Hypertension* 4:686.

Staessen, J., Bulpitt, C., Fogard, R., Joossens, J. V., Lijnen, P., and Amery, A., 1983, Four urinary cations and blood pressure, *Am. J. Epidemiol.* **117**:676.

Stamler, R., Stamler, J., Riedlinger, W. E., Algera, G., Roberts, R. H., 1978, Weight and blood pressure findings in hypertension screening of 1 million Americans, *JAMA* **240**:1607.

Stamler, R., Stamler, J., Grimm, R., Gosch, F. C., Elmer, P., Dyer, A., Berman, R., Fishman, J., Van Heel, N., Civinelli, J., and McDonald, A., 1987, Nutritional therapy for high blood pressure: final report of a four-year randomized controlled trial -- The Hypertension Control Program, *JAMA* **257**:1484.

Stern, B., Heyden, S., Miller, D., Latham, G., Klimas, A., and Pilkington, K., 1980, Intervention study in high school students with elevated blood pressures, *Nutr. Metab.* **24**:127.

Strazzullo, P., Siani, A., Guglielmi, S., Di Carlo, A., Galletti, F., Cirillo, M., and Mancini, M., 1986, Controlled trial of long-term oral calcium supplementation in essential hypertension, *Hypertension* **8**:1084.

Svetkey, L. P., Yarger, W. E., Feussner, J. R., DeLong, E., and Klotman, P. E., 1987, Double-blind, placebo-controlled trial of potassium chloride in the treatment of mild hypertension, *Hypertension* **9**:444.

Swales, J. D., 1982, Ion transport in hypertension, *Biosci. Rep.* **2**:967.

Treasure, J., and Ploth, D. I., 1983, Role of dietary potassium in the treatment of hypertension, *Hypertension* **5**:864.

Tuck, M. L., Sowers, J., Dornfeld, L., Kledzik, G., and Maxwell, M., 1981, The effect of weight reduction on blood pressure, plasma renin activity, and plasma aldosterone levels in obese patients, *N. Engl. J. Med.* **304**:930.

Uusitupa, M., Tuomilehto, J., Karttunen, P., and Wolf, E., 1984, Long-term effects of guar gum on metabolic control, serum cholesterol and blood pressure levels in type 2 (non-insulin-dependent) diabetic patients with high blood pressure, *Ann. Clin. Res.* **16** (Suppl. 43):126.

Watt, G. C. M., Edwards, C., Hart, J. T., Hart, M., Walton, P., and Foy, C. J. W., 1983, Dietary sodium restriction for mild hypertension in general practice, *Br. Med. J. (Clin. Res.)* **286**:432.

Watt, G. C. M., Foy, C. J. W., Hart, J. T., Bingham, G., Edwards, C., Hart, M., Thomas, E., and Walton, P., 1985, Dietary sodium and arterial blood pressure: evidence against genetic susceptibility, *Br. Med. J.* **291**:1525.

Wester, P.-O., and Dyckner, T., 1987, Magnesium and hypertension, *J. Am. Coll. Nutr.* **6**:321.

Zoccali, C., Cumming, A. M. M., Hutcheson, M. J., Barnett, P., and Semple, P. F., 1985, Effects of potassium on sodium balance, renin, noradrenaline and arterial pressure, *J. Hypertension* **3**:67.

Zoccali, C., Mallamaci, F., Delfino, D., Ciccarelli, M., Parlongo, S., Iellamo, D., Moscato, D., and Maggiore, Q., 1986, Long-term oral calcium supplementation in essential hypertension: a double-blind, randomized, crossover study, *J. Hypertension* **4** (Suppl. 5):S676.

Chapter 4

Energy Metabolism of the Newborn Infant

Robin K. Whyte and Henry S. Bayley

1. Introduction

For the newborn mammal the burden of homeothermy is great. Birth is accompanied by a rapid increase in the rate of heat loss, which must be matched by heat production if thermal homeostasis is to be maintained. Growth adds to the energy requirement of the neonate in the days following birth. The provision of an intake of energy greater than that lost as heat is critical for the newborn. Failure to do this results in poor weight gain or even, in low-birthweight infants, an increased risk of death.

The diet of the newborn consists entirely of milk or of infant formula made from modified cows' milk (or sometimes from soybean or other vegetable sources) designed to simulate the macronutrient composition of human milk. These milks have an energy content of about 2.9 MJ/liter. The newborn's energy intake is a function of its milk intake, and is limited by the functional capacity of the gastrointestinal tract. In some cases, a limitation of milk intake may result in an inadequate supply of energy resulting in growth failure. On the other hand, an intake of too much energy intake leads to excess fat storage. Fat is a substantial component of the body which varies in size independently of the lean body mass, and this independence allows a chronic excess of energy intake over expenditure to result in obesity.

Robin K. Whyte and Henry S. Bayley • Department of Pediatrics, McMaster University, Hamilton, ON, and the Department of Nutritional Sciences, University of Guelph, Guelph, ON

Advances in Nutritional Research, Vol. 8
Edited by Harold H. Draper
Plenum Press, New York, 1990

2.   The Concept of Energy Balance

The derivation of modern bioenergetic principles has been described by Brody (1945) and Kleiber (1975a). These principles are studied by calorimetry (Kleiber 1975a) and are derived from the First and Second Laws of Thermodynamics. The amount of heat produced in a chemical process is independent of the intermediary steps by which a system changes from an original to a final state. For example, when glucose is metabolized to carbon dioxide and water, it makes no difference to the energy produced by the reaction, 16 kJ/g of glucose, whether it takes place by direct oxidation in glycolysis and the tricarboxylic acid cycle or by the indirect route of lipogenesis followed by fat oxidation. The bomb calorimeter is used to measure the energy content (heat of combustion) of food and excreta, and energy expended by an organism as heat is either measured by direct calorimetry or calculated from respiratory exchange (indirect calorimetry).

A target state of energy balance can be described in which energy intake matches the energy excreted, expended and needed for optimal growth. In the adult in zero energy balance, energy intake equals energy expenditure and excretion. When energy balance is positive, as in the growing infant or in recovery from weight loss, a further component must be added to account for the energy stored as growing tissue. The complete energy balance equation conforms with the principle of conservation of energy and describes the distribution of energy intake:

Gross energy intake =
> Energy stored + Energy transferred to the environment

Energy is transferred ("lost") to the environment either as excreted energy-containing substances, as kinetic energy (generally as heat) or when work is done on the environment to increase its potential energy. The energy transferred to excreted substances includes both the heat of combustion and the heat content of the excreta. The transfer of heat from the body to both the environment and to mechanical work performed on the environment are forms of energy expenditure, so that:

> Energy expenditure = Heat produced + Work done

Sinclair (1978) has applied the general formula for energy balance to the study of growing infants:

Gross energy intake =
> Energy excreted + Energy stored + Energy expended

Direct calorimetry measures the heat flow from the body to the environment, i.e. heat *loss*. Indirect calorimetry allows the estimation of the total heat *production* and work done from measurements of oxygen consumption, carbon dioxide production and nitrogen excretion, and therefore measures total energy expenditure. Direct calorimetry underestimates energy expenditure when there is a change in body temperature (heat storage) or when mechanical work is done on the environment within the measuring chamber. Positive or negative changes in heat storage take place when there is a change in the thermal environment or in heat production, and they result in changes in the distribution of body temperature between the core and the skin. The effect on heat storage may be estimated by combining multiple measurements of body temperature with assumptions about the specific heat of body tissues (Heim, 1984). In a 70 kg man with an energy expenditure of 10 MJ/d, a change in mean body temperature of 0.4° C would be equivalent to 100 kJ or 1% of daily energy expenditure. This is also equivalent to the energy required to raise a 102 kg weight by 100 meters[1].

Measurements in newborn infants are usually made in controlled thermal conditions, and newborns do not ordinarily perform mechanical work on their environments. Thus, unless acute changes are taking place during short-term measurements, the differences in estimates obtained by direct and indirect calorimetry are likely to be very small. However, Sauer *et al.* (1984) have described a 7% difference between heat loss and energy expenditure ("metabolic rate") in simultaneous measurements in growing low birthweight infants. It is not clear how much of this difference can be accounted for by technical factors: the explanation offered that it represents energy expended in synthesis is not consistent with the principles of conservation of energy.

Most current measurements of energy expenditure in infants are made by indirect calorimetry using an open circuit system (Lister *et al.*, 1974). Open circuit indirect calorimetry is a relatively simple and non-invasive technique, but it requires very sensitive instruments to measure small differences in the oxygen and carbon dioxide content of inspired and mixed expired gases. Long term measurements of energy expenditure can be made using the dual isotope technique, which estimates carbon dioxide production rates from turnover rates of deuterated water and H_2O^{18}: this technique has recently been applied to the newborn (Jones *et al.*, 1987; Lucas *et al.*, 1987).

---

[1] The force required to raise a mass of 102 kg against gravity (9.81 meters/sec^2) is 1000 newtons. When raised to a height of 100 meters the mass gains a potential energy of 100 kilojoules.

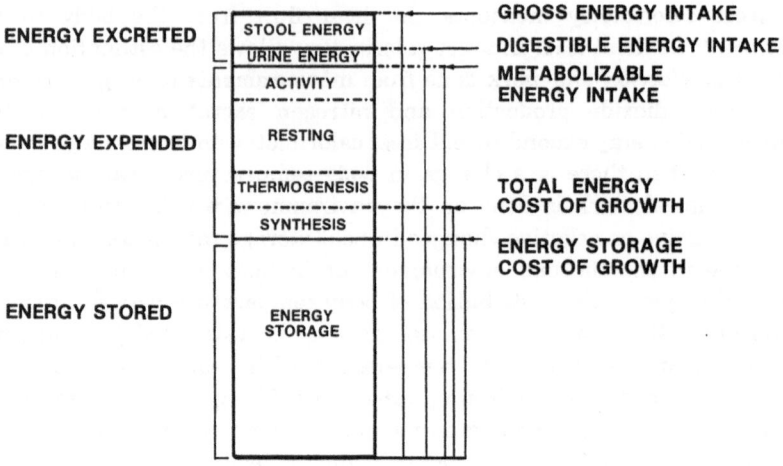

Fig. 1.   Partition of energy intake in growing infants.

3.   The Partition of Energy Intake

Gross energy intake is determined from the total amount and heat of combustion of the diet. The disposition of gross energy intake is illustrated in Fig. 1. Part of the gross energy intake of an infant is lost to the body as energy excreted in the stool; that remaining is called digestible energy intake. Further subtraction of urine energy excretion leaves metabolizable energy intake, or the energy available for metabolism. The energy not expended in metabolism is stored either in growing tissue or, if body weight remains constant, in changes in body composition. Thus the sum of the amounts of energy expended and stored equals the metabolizable energy intake of the infant. Alternative expressions for metabolizable energy intake have been used in the literature, but these will not be used here.

In growing infants, the energy content of growing tissue can be described as the energy storage cost of growth. The combination of energy stored and energy expended in the synthesis of new tissue constitutes the total energy cost of growth.

In infants, gross energy intake is derived from the carbohydrate, protein and fat of milk or formula. The relationship between gross, digestible, and metabolizable energy in the infant is different from that in the adult. Fat makes up between 40 and 50% of the energy content of a milk diet. The term newborn infant absorbs 85-90% (Fomon *et al.*, 1970) and the preterm infant 65-70% of formula fat (Tantibhedhyankul *et al.*, 1975). In contrast, the adult absorbs some 95% of dietary fat. These differences in fat absorption are due largely to the relative small quantities of bile salts and pancreatic lipase secreted by the newborn infant (although much of neonatal fat digestion takes place within the stomach from the action of lingual lipase (Hamosh, 1979). Chappell *et al.* (1986) have reported higher coefficients of fat absorption for the preterm infant; 80% for formula fat and 90% for freshly expressed mother's milk. This may reflect improved formulation of the former and the action of breast milk lipases in the latter (Hamosh, 1982). The absorption of carbohydrate is nearly complete in newborn infants, although a large part of the disappearance of carbohydrate from the intestine is due to fermentation by the intestinal flora (Kien *et al.*, 1982; MacLean *et al.*, 1980).

It is common practice to estimate the energy content of foods by multiplying the gram concentrations of their fat, protein, and carbohydrate contents by Atwater's conversion factors (37.5, 17, and 16 kJ/g, respectively). These values are *fuel values* determined in healthy young adults rather than *heats of combustion* (39, 24 and 17 kJ/g, respectively) (see data of Atwater reviewed by Merrill and Watt, 1973). When these values are used to estimate the energy content of infant formula, they result in an underestimate of gross energy content (heat of combustion) and an overestimate of metabolizable energy content (fuel value) of formula fed to infants (Table I). Calculation of the gross energy content of milk and formula from protein, fat, and carbohydrate content using Atwater's values for heats of combustion for milk protein, fat, and carbohydrate gives values very close to those obtained by bomb calorimetry (Lemons *et al.*, 1982).

Most of the differences in formula digestibilities between preterm infants, term infants and adults reflect differences in fat absorption. The difference between energy digestibility and metabolizability is small, and depends on energy excreted in urine, chiefly in the form of urea. Nitrogen excretion, chiefly as urea, is related to the amount of protein digested and to nitrogen retention in growing tissue. In the adult and the non-growing infant, urea nitrogen excretion matches nitrogen intake from digested nitrogenous products. Thus the difference between digestible and metabolizable energy intake is affected by protein intake and the rate of growth. This is, however, a small component of energy balance; in a preterm infant ingesting 2.4 g protein per day, a change in nitrogen retention from 70% to zero would result in an increase in urine energy excretion from 2 to 6 kJ/d, a reduction of only 1% in metabolizable energy content.

Table I. Relationship between Gross, Digestible and Metabolizable Energy Contents of Infant Formula (MJ/liter) in Preterm Infants[a], Term Infants at One and Six Months of Age[b] and Adults[c]

Energy Content of Formula	Preterm Infant	Term Infant 1 mo	Term Infant 6 mo	Adult
		MJ/Liter		
Gross	3.00	3.00	3.00	3.00
Digestible	2.61	2.78	2.82	2.88
Metabolizable	2.58	2.75	2.79	2.82[*]
% of Digestible	87	93	94	96
Metabolizability	86	92	93	94

[*] Value used in label declaration

The data represent expected values for energy content of a liter of formula containing 150 g protein, 35 g fat and 75 g lactose.

[a] Whyte et al. (1983)   [b] Fomon et al. (1970) and Fomon (1974)

[c] Data for dairy foods from Merrill and Watt (1973)

4.   Energy Expenditure of the Fetus and Infant

4.1.   Energy Expenditure of the Fetus

Energy metabolism in the human fetus is not readily determined by direct physiological measurement. The data available are confined to measurements on cord blood at the time of delivery, speculations from pathological material, or observations on the prematurely born. There is a great deal of literature on the fetal physiology of other species, reviewed by Silver (1976) and Battaglia and Meschia (1978), in particular that of the sheep and the horse. Current estimates of fetal oxygen consumption in animals are fairly uniform (between 6 and 9 ml·kg^{-1}min^{-1}). During rapid fetal growth substantial amounts of carbohydrate are converted to fat, a carbon dioxide generating process. The respiratory quotient is used to derive the thermal equivalent of oxygen consumed, on the assumption that all carbon dioxide is derived from oxidation. The production of carbon dioxide from lipogenesis makes it difficult to assign a thermal equivalent for oxygen consumption. Based on a calculation of the likely metabolic fuel mixture of amino acids and glucose, Battaglia and Meschia (1978) assigned a thermal equivalent of 21 kJ/liter for oxygen consumed by the fetal lamb.

Differences in fetal body composition and rates of growth, and in placental metabolism and architecture, make extrapolation of data for the fetuses of non-human species to the human fetus of limited value. Sinclair (1976) combined estimates of oxygen consumption of the human fetus (Romney et al. (1955) and of human placental tissue (Villee, 1953) to estimate the oxygen consumption of the human feto-placental unit to be about 5 ml·kg^{-1}min^{-1} (150 kJ·kg^{-1}d^{-1}). Sparks et al. (1980) have argued that the rate of oxygen consumption of the human fetus is likely to bear a relationship to fetal weight similar to that of the simian fetus, and therefore is more likely to be 8 ml·kg^{-1}min^{-1}(240 kJ·kg^{-1}d^{-1}).

4.2   Energy Expenditure at Birth

Birth is characterized by a temporary period of hypoxia, an interruption of the placental supply of nutrients, and a reduction of ambient temperature. There is an abrupt reversal of anabolic pathways to provide substrate for energy metabolism (Adam, 1971). The rapid fall in blood glucose, the adrenergic response to birth and the post-neonatal surge of glucagon are thought to be factors in the activation of gluconeogenic pathways, particularly through their influence on phosphoenolpyruvate carboxykinase, the transaminases and 2,3-diphosphoglycerokinase (Ferre, 1986). As glycogen stores become exhausted, and oxygen becomes available, a predominantly glycolytic form of energy expenditure is replaced by fat metabolism (Persson and Gentz, 1966), reflected in a fall in the respiratory energy equivalent to 0.7 (Senterre and Karlberg, 1970). The first feeds of life are associated with transient post-prandial rises in the respiratory energy equivalent (Gentz et al., 1976), and once growth is established it remains high (0.85-0.98), reflecting the importance of carbohydrate as a metabolic fuel.

4.3   Energy Expenditure and its Variables in the Neonate

The rate of energy expenditure is highly variable. The normal newborn infant is never fasting, but is either feeding, post-prandial (and often sleeping) or exhibiting the increasing activity associated with expectation of the next meal. Thus, there can be no measurement of "basal metabolic rate" as defined for the resting, fasted adult. Many measurements of infant metabolic rate are confined to short measurements of "resting" metabolic rate, usually described as post-prandial measurements during sleep or minimal physical activity outside the postprandial period. Many of the earlier measurements of infant metabolic rate were conducted in conditions which were incompletely defined with respect to many of the important experimental variables, in particular those describing feeding patterns and the thermal environment. These have been critically reviewed by Sinclair (1976).

Total energy expenditure over a 24-hr period or longer has hitherto only rarely been measured in newborn infants. Bell *et al.* (1986) and Abdulrazzaq and Brooke (1984) have shown that in low birthweight infants acceptable levels of accuracy for estimation of 24-hr energy expenditure require measurement for at least 6 hr or two full inter-feeding epochs. Energy expenditure is controlled by the action of thyroid hormones, a slowly responding system affecting obligatory thermogenesis, and the sympathetic nervous system which responds immediately to stimuli such as acute thermal stress, energy intake and exercise (Weinstein *et al.*, 1987). The variables affecting energy expenditure are body weight, postnatal and gestational age, physical activity, the thermal environment, the effects of feeding, the rate of weight gain, and the state of health of the infant. The effects of these variables are discussed below.

The literature on energy expenditure contains data expressed as heat production or as oxygen consumption. For the purposes of comparison, results will be expressed as units of energy expenditure ($kJ \cdot kg^{-1} d^{-1}$). When data are reported as oxygen consumption, the R.Q. will be used to derive the thermal equivalent of oxygen consumed, or, when the R.Q. is not given, a thermal equivalent of 20.3 kJ/liter will be used (Karlberg, 1952). The thermal equivalent ranges from 19.6 kJ/liter at R.Q. 0.7 to 21.1 kJ/liter at R.Q. 1.0 (Brody, 1945), so the maximal error introduced by using the constant ranges from -3% to +2.0%. The maximal error introduced by calculating energy expenditure from the overall R.Q. rather than from the combined non-protein R.Q. and urinary nitrogen excretion ranges from +0.96% at an R.Q. of 0.95 to +0.83% at an R.Q of 0.75, and can for most purposes be ignored.

4.3.1. Effect of Body Weight

A unifying theory relates the metabolic rate of different animals to the three-fourth power of body weight: thus among adult homeothermic animals of different size, the value for metabolic rate in $kJ \cdot kg^{0.75}$ is fairly constant (Kleiber, 1975b). This relationship, developed for adult animals of different species, does not apply to human newborns of different body weight: paradoxically, the relationship is best described by an exponent for weight greater than one, so that heavier infants have greater metabolic rates per unit body weight (Sinclair et al., 1967). Most recent expressions of metabolic rate have used a simple relationship with body weight (kJ per kg), and this form of expression will be used here.

4.3.2. Effect of Postnatal and Gestational Age

The energy expenditure of the term newborn infant for the first few days of life is relatively low (80 - 100 $kJ \cdot kg^{-1}d^{-1}$) (Senterre et al., 1970) and rises over the first week of life to about 180 $kJ \cdot kg^{-1}d^{-1}$. Bhakoo and Scopes (1970) demonstrated a rise in minimal rates of energy expenditure of small-for-dates infants from 106 to 138 $J \cdot kg^{-1}min^{-1}$ over the first three days of life, and compared this to the much slower rise seen in comparable infants in their studies carried out four years before, in which infants took eight days to experience a rise of similar magnitude. They attributed the increased rate of rise in the later study to a change in feeding practice; in the earlier study, feedings were introduced very slowly, reaching a gross energy intake of 180 $kJ \cdot kg^{-1}d^{-1}$ on the third day of life, compared to an intake of 360 $kJ \cdot kg^{-1}d^{-1}$ experienced by infants in the later study. A positive relationship between feeding and energy expenditure in newborn piglets was reported by Gentz and Kellum (1971). An effect of postnatal age on energy expenditure independent of increasing energy intake has not been demonstrated.

Estimates of energy expenditure of healthy full birthweight breast-fed infants over the first twelve weeks of life have been made using the dual isotope technique (Lucas et al., 1987). There was no significant difference in energy expenditure with postnatal age, and expenditures were 280 and 300 $kJ \cdot kg^{-1}d^{-1}$ at five and eleven weeks, respectively. These values are much higher than the 230 $kJ \cdot kg^{-1}d^{-1}$ reported for resting energy expenditure by Benedict and Talbot in 1915, which hitherto have been used as a standard for this age. The dual isotope technique, if it is adequately validated, is more applicable than indirect calorimetry for long term studies of normal infants; the higher values reported for energy expenditure may reflect the measurement of total rather than resting values.

The hospital confinement of preterm infants has made them more ready subjects for studies at later postnatal ages than infants of normal birthweight. A very large post-natal rise in energy expenditure from 180 to 270 $kJ \cdot kg^{-1}d^{-1}$ from the fifth to the fifteenth day has been described for preterm infants by Chessex et al. (1981), and as this rise was accompanied by an increase in gross energy intake from 330 to 650 $kJ \cdot kg^{-1}d^{-1}$, the postnatal rise in energy expenditure was attributed to the increase in energy intake. Hill and Robinson (1968) described a gradual increase in minimal rates of energy metabolism in low birthweight infants from 200 $kJ \cdot kg^{-1}d^{-1}$ at one week to 260 $kJ \cdot kg^{-1}d^{-1}$ at two months. Longitudinal studies of low birthweight infants over the first ten weeks of life (Sinclair, 1976) show a gradual increase in minimal energy expenditure from 200 to 250 $kJ \cdot kg^{-1}d^{-1}$. More recent measurements of the total daily energy expenditure of low birthweight infants in the first month of life (220-250 $kJ \cdot kg^{-1}d^{-1}$) (Whyte et al., 1983; Reichman, 1981) are consistent with these values. Thus it appears that low birthweight infants, unlike term infants, experience a gradual increase in energy metabolism over the first months of life. It is unclear whether or not this increase is related to postnatal age, increasing energy intake or to changing body composition.

4.3.3.   Effects of Physical Activity and Behavioral State

These variables are related to feeding, and it is difficult to isolate their individual effects. Most investigators have addressed this problem by measuring energy expenditure in certain predefined behavioral states before and after feeds. The effects of activity were described by Murlin et al. (1925) in a study of fifty children carried out immediately following feeds in an incompletely defined thermal environment. Crying was reported to double the rate of energy expenditure measured in sleeping, post-prandial infants. Only one form of sleep was defined: newborn infants spend 15 to 18 hr a day asleep (Schulte, 1981). "Active" or rapid-eye-movement sleep occupies 50% of sleeping time and even more in the preterm infant, and energy expenditure during active sleep is 5% to 7% greater in both term and preterm infants than it is during quiet sleep (Stothers and Warner, 1984; Stothers and Warner, 1979). Physical activity of term newborns contributes only 10% to the total energy expenditure (Stothers and Warner, 1977).

The association of activity with energy expenditure has been described for preterm infants. Brooke et al. (1979) reported an increase in energy expenditure as activity increased from sleep (140 $J \cdot kg^{-1}min^{-1}$) through quiet awake (175 $J \cdot kg^{-1}min^{-1}$), active awake (280 $J \cdot kg^{-1}min^{-1}$) and crying states (330 $J \cdot kg^{-1}min^{-1}$). Rubecz and Mestyan (1975) reported that, as in term infants, preterm infants in a thermoneutral environment were largely inactive, so that physical activity contributed only 10% to daily energy expenditure. For any given activity the rate of energy expenditure on a body weight basis increased with age, reflecting the increase in muscle mass. Mestyan *et al.* (1968) observed a large increase in physical activity in cold environments: half of the increase in energy expenditure associated with exposure to cold was attributable to increased activity.

4.3.4.   Effect of the Thermal Environment

The newborn infant has a high surface area to weight ratio and is particularly susceptible to heat loss. Although the effect of ambient temperature on the metabolism of newborn infants and animals was previously recognized (Herrington, 1940; Day, 1943; Hill, 1959), the mechanism and magnitude of the response in low birthweight infants was first described by Bruck *et al.* (1961). Radiant and evaporative losses form substantial components of total heat loss: for example, in a thermoneutral environment of uniform temperature (34° C) and 50% humidity, the partition of heat loss by a naked 4-day-old premature infant weighing 1.5 kg is 34% convective, 39% radiant, 24% evaporative and 3% conductive (Hey, 1973). The rate of radiant heat loss depends upon surrounding wall temperatures and that of evaporative loss upon humidity (Hey and Mount, 1967).

If the routes of radiant and evaporative heat loss are held constant, the thermal environment can be described in terms of ambient temperature: within a certain range of temperature there is no change in heat production (Hey, 1970). Thermal homeostasis is maintained by peripheral vasoconstriction and vasodilation (Day, 1943) and by postural change (Stothers and Warner, 1984). At the upper limit of the neutral thermal environment, evaporative heat loss increases abruptly when the infant sweats (Hey and Katz, 1969). The zone between the critical temperatures at which energy expenditure is minimal is known as the zone of minimal heat production: it is wider than the thermoneutral zone, whose upper limit is defined by the point at which evaporative heat loss increases. A detailed discussion of these definitions and of the mechanisms of heat loss of the newborn is given by Swyer (1978).

When the heat losing properties of the environment exceed those of the neutral thermal environment (i.e. in "cold" environments), energy expenditure increases (Hey, 1969). The increment may more than double the rate of energy expenditure. In the first week of life, however, and in preterm infants, the maximal response is less, and the lower limit of the neutral thermal environment is higher. Unlike the adult response to cold, the increase in energy expenditure takes place largely without shivering (Day, 1943; Hey, 1969). It is generated instead by an increase in physical activity (Mestyan, 1968) and the oxidation of fat by brown adipose tissue. Brown adipose tissue is abundant in the newborn infant, in which it is distributed around the neck, between the scapulae and around the spinal cord, where venous drainage is thought to specifically favor transfer of heat to the spinal cord. The metabolism and function of brown fat have been reviewed extensively (Hull, 1966; Sinclair, 1976; Benito, 1985).

In "warm" environments, energy expenditure increases due to active sweating and to increases in cardiac output and skin blood flow. As these activities themselves increase heat production, homeostatic control may easily be lost; the newborn infant is therefore very vulnerable to heat exhaustion. Once thermal homeostasis is lost, energy expenditure is further increased by the Vant Hoff effect on metabolism. Unless the thermal environmental temperature is rapidly corrected, heat production and deep body temperature will continue to increase exponentially until death occurs.

The importance of the thermal environment for low birthweight infants was confirmed by Silverman et al. (1958), who demonstrated a significantly increased mortality (19% vs 7%) in low birthweight infants nursed at a cooler incubator temperature (29° C vs 32° C). A further series of clinical trials (reviewed by Hey, 1973) demonstrated additional improvements in survival with more closely controlled adjustment of the thermal environment. The effect of the increase in energy expenditure in cold environments was shown by Glass et al. (1968) to be reflected in a reduced rate of weight gain. Clear definitions of the thermal environment for normal and low birthweight naked and clothed infants at various postnatal ages were developed (Hey and Katz, 1970; Hey and O'Connell, 1970). These guidelines have now been generally adopted for incubator control of the thermal environment. Further improvements came with the introduction of servo-control of air temperature by skin temperature (Buetow et al., 1964) for both incubators (Bell and Rios, 1983) and radiant warmers (Malin and Baumgart, 1987), and with the use of proportional heating system responses (as opposed to on-off switching). Devices have been described which reduce radiant and evaporative heat loss, such as heat shields (Baumgart et al., 1981) and double walled incubators (Perlstein, 1976), and there has been a reevaluation of different forms of clothing for reducing heat loss, particularly from the head (Marks, 1977; Stothers, 1981).

4.3.5.   Effects of Feeding and Weight Gain

The response of the term infant to feeding has not been clearly documented. In two term infants, Levine et al. (1927) recorded a 15-17% increase in energy expenditure in response to a protein meal, and a much smaller response to meals of fat and carbohydrate. Stothers and Warner (1979) were not able to identify a postprandial effect of feeding in nine term infants, but did not begin measurements until one hour after the feed. Brooke and Alvear (1982) did not find a significant rise in postprandial metabolism in 15 term infants.

The effect of feeding has, however, been clearly documented in preterm infants. Mestyan *et al.* (1969), using feeds of high protein content (39 g protein and 2.9 MJ gross energy per liter), described increases in energy expenditure averaging 9% above total daily expenditure (or 26% above resting expenditure). The response was maintained even when energy expenditure was already increased by thermal stress. Stothers and Warner (1979), in measurements confined to periods of rapid-eye-movement sleep, described a mean rise of 13% over resting values in nine preterm infants fed mother's milk or formula. The response lasted from 15 to 45 min after the feed. Brooke and Alvear (1982) described a rise of similar duration and magnitude (17%) in low birthweight infants, but found a smaller percentage rise in infants who were small for gestational age; they attributed this surprising finding to the high resting metabolic rate found in the latter group. Freymond *et al.* (1986) have estimated the separate effects of physical activity and feeding in preterm infants by conducting linear regression analysis of the effects of activity on energy expenditure both before and after feeding. Premeal activity was higher than postmeal activity, and the energy expenditure attributable to this was taken into account. Assuming that the measuring period included all of the post-prandial effect, only 13 kJ·kg^{-1}d^{-1} of a total expenditure of 286 kJ·kg^{-1}d^{-1} was attributable to the feed, which accounted for 3% of metabolizable energy intake. Energy expenditure attributable to activity was 26 kJ·kg^{-1}d^{-1} before feeds and 19 kJ·kg^{-1}d^{-1} after feeds. In measurements made in infants feeding frequently (Whyte *et al.*, 1983), feeding may start in a post-prandial period, making the effect undetectable.

The biological mechanism by which feeding results in an increase in energy expenditure involves the energy costs of digestion and assimilation of a feed (Armsby, 1917). The theoretical estimate of the energy expended in these processes (referred to variously as heat increment, specific dynamic action and thermal energy) for various meals closely predicts values obtained experimentally in adults (Baldwin and Smith, 1974). The synthesis of a gram of protein from amino acids and a gram of fat from acetate can be calculated to require the expenditure of 4.3 and 11.9 kJ, respectively. Energy is expended in the synthesis and breakdown of fat and protein, and both these processes contribute to postprandial energy expenditure. Growth is a process of exchange, in which the rate of synthesis exceeds the rate of breakdown of tissue components. The rate of energy expenditure attributable to growth can be calculated only from knowledge of both the rate of synthesis and of breakdown, and therefore in the absence of estimates of turnover rates the energy expended in growth cannot be calculated from net gain alone. However, Pullar and Webster (1977) have determined the energy cost of deposition of fat and protein in growing rats with gains of different composition (i.e., growing obese rats and growing lean rats). Energy costs were determined   at different body weights and energy intakes so as to obtain data for eight different combinations of growth of fat and protein. Multiple regression analysis was used to predict the energy expended in the net deposition of fat and protein. For each kJ stored as protein 2.25 kJ of metabolizable energy was required, compared to 1.36 kJ of metabolizable energy for each kJ stored as fat. Thus for every gram of protein deposited (24 kJ), 30 kJ are expended, while for every gram of fat deposited (39 kJ), 14 kJ are expended. Hence about the same metabolizable energy (54 and 53 kJ) is required to deposit a gram of either protein or fat. Measurements of the energy expenditure cost of growth have been attempted in low birth-weight infants by examining the relationship between energy expenditure and weight gain; values calculated in this way were 2.3 kJ (Gudinchet *et al.*, 1982) and 5.7 kJ (Brooke and Alvear, 1982) for each gram of growth.

It is not clear whether all the increased energy expenditure attributable to growth is limited to the postprandial period. As resting energy expenditure is not measured in true basal (starving) conditions in the newborn, the true thermic effect of food may be underestimated. Resting energy expenditure was positively related to weight gain in the study of Brooke *et al.* (1979). The effect of metabolizable energy intake is most clearly seen when intravenous energy intakes are varied (Weinstein *et al.*, 1987). Low birthweight infants in the first week of life, randomized to receive either 159 or 268 $kJ \cdot kg^{-1} d^{-1}$ intravenously, had energy expenditures of 152 and 196 $kJ \cdot kg^{-1} d^{-1}$ respectively. The increase in energy intake was accompanied by a four-fold increase in urinary norepinephrine excretion, reflecting mediation of this response through sympathetic control.

4.3.6.   Effect of Body Composition, Health and Oxygenation

The oxygen consuming component of the body is known as the body cell mass. Energy expenditure per unit body weight of preterm infants increases with gestational age, and this is a reflection of the proportional increase in the body cell mass of the fetus in the last trimester (Sinclair and Silverman, 1966). Infants with intrauterine growth retardation have, on average, 17% higher rates of oxygen consumption per unit body weight than have infants of similar birthweight who have appropriate birthweight for gestational age, and this is presumably a reflection of an increased cell mass to body weight ratio (Sinclair and Silverman, 1966; Sinclair *et al.*, 1967).

There have been very few studies of energy metabolism performed on sick infants. In mechanically ventilated infants with hyaline membrane disease, the rate of oxygen consumption at the height of the disease is about 60% above recovery levels (Richardson *et al.*, 1984). In infants with the chronic lung disease bronchopulmonary dysplasia, oxygen consumption is raised by 25% compared to healthy infants (Weinstein *et al.*, 1981). The growth of infants with this disease is often very poor and is inversely related to energy expenditure, which in turn is related to the increased work of breathing caused by the disease (Kurztner et al., 1988). Catch-up growth characteristically takes place at the time of resolution of the respiratory disease (Markestad and Fitzharding, 1981). This is probably a reflection of the effect of increased energy expenditure on energy storage and weight gain. There is an increase in the rate of oxygen consumption in infants with cyanotic congenital heart disease. In anencephaly, the resting rate of oxygen consumption is reduced and there is a reduced or absent thermal response to cooling (Cross et al., 1966).

Oxygen consumption can be limited by the rate of oxygen supply to the tissues. Although little is known about the way in which reduced tissue oxygenation limits resting energy expenditure in humans, the increase in oxygen consumption of low birthweight infants in response to cold stress can be abolished by hypoxia (Scopes and Ahmed, 1966). The point at which systemic oxygen supply limits the rate of resting oxygen consumption is higher in the newborn than in the adult sheep, reflecting the higher resting values of systemic oxygen consumption of the newborn animal (Moss et al., 1987).

5.   Growth and Energy Storage in the Fetus and Infant

Energy stored in growing tissue accounts for a major portion of energy intake during the rapid growth phases of the fetus and infant. The energy storage cost of weight gain can be expressed as kilojoules stored per gram of weight gain. A high value suggests that much of the weight gain consists of fat and a low value that much consists of water. Protein accretion (and its equivalent as energy) can be calculated from nitrogen retention and, given that carbohydrate storage is negligible (Widdowson and Dickerson, 1964), the remaining stored energy can be assumed to consist of fat. Thus the composition of weight gain can be described in percentages of fat and protein, the remaining weight gain consisting of water and minerals.

There are few reliable estimates of fetal weight prior to the third trimester and fewer assessments of body composition. Data derived from cross-sectional analysis has indicated that fetal weight gain proceeds at a constant rate in midpregancy and slows toward the end of pregnancy (McKeown, 1952; Usher, 1969; Kloosterman, 1970; Sinclair, 1984). This has been interpreted as reflecting the limitation of nutrient supply via the placenta, although the effect of mechanical restriction of the fetus by the uterus may also be a factor. Such estimates of fetal weight gain have generally been made by measuring the weight of the newborn infant at various lengths of gestation: these techniques are liable to underestimate

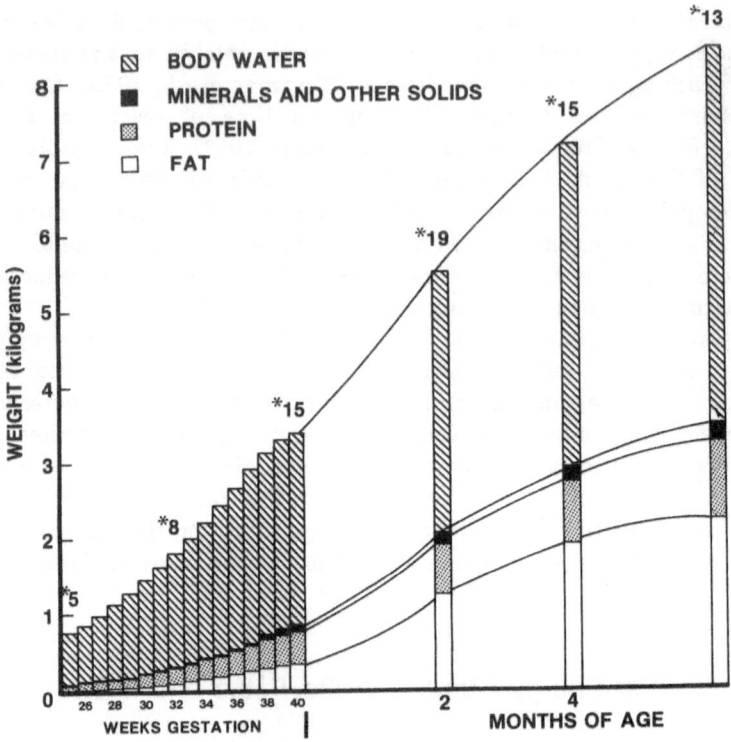

Fig. 2. Growth and body composition of the fetus and infant. The weight of the body and its components is shown for the fetus up to 40 weeks gestation (Ziegler, 1976) and for the infant up to 6 months of age (Fomon, 1967).

healthy fetal weights before term, given that premature birth is a pathological event (Tanner, 1970). In studies in which rigorous attention is given to avoiding selection bias (Sinclair *et al.*, 1982), distortion can still occur as a result of a biological effect of fetal size on the onset of labor, and thus on gestational age. Such weight curves are therefore correctly referred to as "size-at-birth" distributions rather than "fetal growth curves." The development of ultrasound has recently allowed true longitudinal estimates of fetal size to be made (Persson *et al.*, 1986), and it appears from preliminary data that the rate of fetal weight gain does not slow down as term approaches.

Assessment of the body composition of the fetus in the third trimester has also depended entirely on cross-sectional data derived from pathological material. Models of fetal growth have been assembled by Widdowson and Dickerson (1964) and by Ziegler *et al.* (1976). Data from the model of Ziegler *et al.* (1976) have been combined in Fig. 2 with data from a model developed for infants from *in vivo* measurements of body composition by Fomon (1967). The water content of the body falls from 88% at 25 weeks gestation to 74% at term, and reaches 59% by one year. Protein content increases from 9% at 25 weeks gestation to 12% at birth and to 15% by the end of the first year. The amount of fat in the body shows the most extreme changes: whole body fat accounts for less than 1% of the body weight before 25 weeks gestation, but accumulates rapidly to reach 11% at birth and a maximum of 26% at four months of age, then declines to 24% at one year of age. There is some support for these estimates given by measurements of the change in skinfold thickness in the first year of life (Hutchinson-Smith, 1973).

The energy stored in the growing tissue reflects these changes in composition, increasing from 5 kJ/g at 25 weeks gestation to 10 kJ/g at term, reaching a maximum in the first four months of approximately 18 kJ/g (Fig. 2). This value then declines progressively over the remainder of the first year of life. Sparks *et al.* (1980) have combined fetal growth and body composition data with estimates of fetal energy expenditure to calculate total energy requirements for the human fetus of 420 kJ·$kg^{-1}d^{-1}$ at term.

Few data exist for direct measurements of the energy balance of term infants; however, a number of studies have been carried out to examine the partition of energy intake into energy excretion, expenditure and storage in low birthweight infants. These allowed an estimate of the energy storage cost of growth of low birthweight infants and a comparison with that of the fetus and newborn infant. The studies included in this summary are confined to those in which the three components of energy balance (excretion, expenditure and storage) were measured, although they include some in which the energy content of the feed and excreta was estimated from nutrient content (Reichman *et al.*, 1983; Putet *et al.*, 1984; Putet *et al.*, 1987; Schulze *et al.*, 1987) as opposed to measurements by bomb calorimetry (Brooke *et al.*, 1979; Whyte *et al.*, 1983; Whyte *et al.*, 1986; Freymond *et al.*, 1986).

The partition of energy intake observed in different studies is remarkably similar despite differences in energy intakes and diets (Fig. 3). An exception is the study of Brooke (1979) in which very large energy intakes (mean 757 kJ·$kg^{-1}d^{-1}$) were reported: this was accompanied by an excretion of 22% of energy intake. It is known that energy excretion increases with large energy intakes (Brooke *et al.*, 1980); otherwise, energy excretion accounts for 6% to 15% of gross energy intake. Metabolizable energy intake is almost equally divided between expenditure and storage, the percentage for expenditure ranging from 48 to 62%. Energy expenditure as well as energy storage has been shown to increase with energy intake (Brooke, 1980; Schulze *et al.*, 1987), and this effect can be seen when studies are compared (Whyte *et al.*, 1985). The three studies in which energy expenditure accounted for more than 60% of metabolizable energy intake are that of Brooke *et al.* (1979), in which the highest energy intakes were offered, that of Putet *et al.* (1987), in which a high protein intake was combined with a relatively low energy intake, and that of Freymond *et al.* (1986) in which the thermal environment may not have been neutral because of experimental conditions (Gudinchet, 1982).

In most studies the energy balance technique was combined with the determination of nitrogen balance in an attempt to estimate the composition

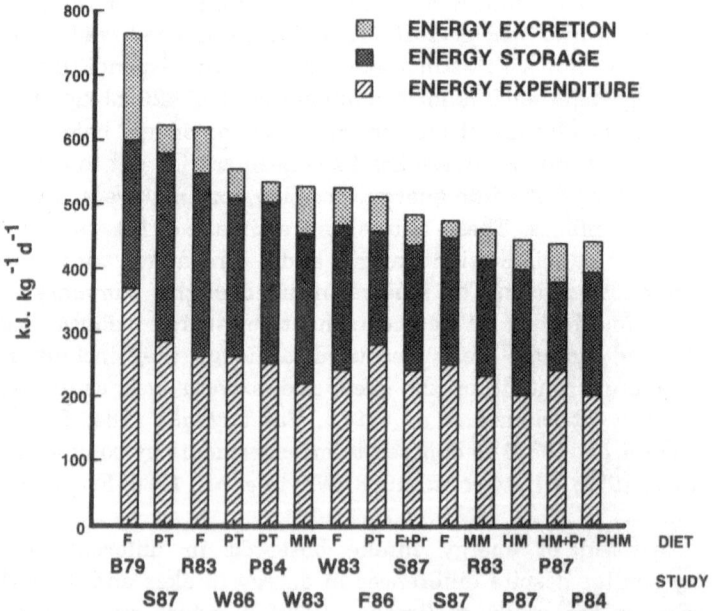

Fig. 3. Energy balance studies in growing low birthweight infants. F: Standard formula; PT: Preterm formula; MM: Own mother's expressed milk; HM: Pooled banked human milk; +Pr: with protein supplementation. Data from B79: Brooke *et al.*,1979; R83: Reichman *et al.*,1983; W86: Whyte *et al.*, 1986; P84: Putet *et al.*,1984; W83: Whyte *et al.*, 1983; F86: Freymond *et al.*, 1986; S87: Schulze *et al.*, 1987; P87: Putet *et al.*, 1987.

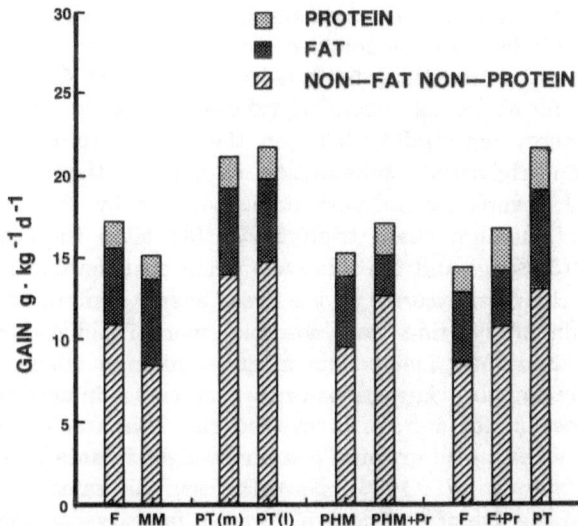

Fig. 4. Composition of weight gain in growing low birthweight infants. F: Standard formula; PT: Preterm formula; PT(m): Preterm formula with medium-chain triglyceride; PT(1): Preterm formula with long-chain triglyceride; MM: Own mother's expressed milk; PHM: Pooled banked human milk; +Pr: with protein supplementation.

of growth of the preterm infant (Fig. 4). In babies with moderate energy intakes (530 kJ·kg⁻¹d⁻¹) from their own mother's milk, the amount of energy stored per gram of growing tissue in the preterm infant (15 kJ/g) exceeds the value for the fetal model of similar gestational age (8 kJ/g) and approaches that of the term newborn (18 kJ/g) (Whyte *et al.*, 1983). With similar intakes of standard formula (2.9 MJ gross energy/liter), the energy stored per gram of growing tissue was lower (13 kJ/g), indicating a slightly reduced accretion of fat in growing tissue. Although significantly different statistically, the clinical consequences of this difference would seem to be small.

The combined energy and nitrogen balance technique has been used to evaluate different feeding regimens. In an attempt to increase weight gain of formula-fed infants, "premature" formulas were introduced, i.e., formulas in which protein, energy and mineral contents were all higher than in human milk-based standards. Clinical trials to study fat digestion indicated that the medium chain triglycerides were almost completely digested (>95%) in the low birthweight infant (Tantibhedyankhul and Hashim 1975; Okamoto *et al.*, 1982) and, as a result, several formulas with partial (12%-50%)

replacement of long chain with medium chain triglycerides were marketed. Whyte *et al..* (1986), in a randomized crossover trial, compared the energy balances of low birthweight infants fed a "premature" formula containing 46% vs 4% of fat as medium chain triglyceride. There was no detectable difference in energy digestibility between the two formulas. The failure of medium chain triglyceride substitution to improve the digestible energy value of the formula is in part accounted for by the lower heat of combustion of medium chain triglycerides (34 kJ/g) than of long chain triglycerides (39 kJ/g) and the relatively small contribution (about 20%) of the medium chain triglyceride to the gross energy content of the formula. Medium chain triglycerides are absorbed more rapidly than long-chain triglycerides, and their metabolism is quite different (Bach *et al.,* 1982). Dietary medium chain triglycerides cause an increase in energy expenditure and a decrease in fat storage in overfed rats (Baba *et al.,* 1982). These effects were not found in growing low birthweight infants by either Brooke (1980) or Whyte *et al.* (1986). Nevertheless, the rates of weight gain reported with the use of "premature" formula far exceeded those seen with standard feeds, and this effect has been consistently observed with formulas of similar composition (Putet *et al.,* 1984; Schulze *et al.,* 1987, Fig. 4). Much of the increase in weight gain obtained with "premature" formulas is attributable to an increased gain of water and minerals. How much of the increment in body water reflects growth of the body cell mass and how much reflects growth of the extracellular space is not clear.

Schulze *et al.* (1987) have examined the energy and nitrogen balances of 19 infants randomly assigned to different intakes of energy and nitrogen. Eight infants were fed with a standard gross energy and protein intake (470 $kJ \cdot kg^{-1}d^{-1}$ and 2.24 $g \cdot kg^{-1}d^{-1}$), and 5 infants were fed with the same amount of energy but with a higher protein intake (3.6 $g \cdot kg^{-1}d^{-1}$). There was little effect on energy balance (Fig. 3) and only a small increment in weight gain (17 vs 14 $g \cdot kg^{-1}d^{-1}$) (Fig. 4) attributable to the protein supplementation; most of the increase in gain consisted of protein and water. In the third group of 6 infants, both energy and nitrogen intakes were increased to 623 $kJ \cdot kg^{-1}d^{-1}$ and 3.5 $g \cdot kg^{-1}d^{-1}$. There was a further increase in weight gain to 21 $g \cdot kg^{-1}d^{-1}$ which was due to the accretion of protein and fat in growing tissue. There was also a greater energy expenditure in this group.

Different findings were reported by Putet *et al.* (1987), who compared eight infants fed pooled human milk at 447 kJ gross energy $\cdot kg^{-1}d^{-1}$ and 2.5 $g \cdot kg^{-1}d^{-1}$ of protein with eight infants supplemented with protein to receive a total of 3.8 $g \cdot kg^{-1}d^{-1}$. The protein supplemented group experienced a 19% increment in energy expenditure at the cost of energy storage, and this increase was reflected in a lower energy and fat content of the weight gain. The difference in rates of weight gain was small (17 vs 15 $g \cdot kg^{-1}d^{-1}$) and did not reach statistical significance.

The interpretation of these studies is limited by small numbers, but a unifying explanation may be attempted. In the study of Schulze *et al.* (1987) the metabolizable energy intake of the groups (means 440-570 kJ·kg⁻¹d⁻¹) was higher than that recorded by Putet (1987) (380-400 kJ·kg⁻¹d⁻¹). When energy intake exceeds the level required to support growth, supplementation with protein may result in an increase in weight gain. When energy intake is low and limits growth, protein supplementation results in increased nitrogen excretion and energy expenditure. Combined supplementation of protein and energy allows further weight gain, until the intake of another nutrient (such as phosphate or sodium) has become limiting. In the studies of Schulze *et al.* (1987) the infants were fed with the high intakes of calcium, phosphorus and minerals typical of a "premature" formula. That the composition of weight gain represented an increase of tissue (as opposed to water) is indicated not only by the relatively low accretion of non-protein non-fat weight (Fig. 4) but also in the accompanying high retentions of phosphorus and minerals described for the same study subjects by Kashyap *et al.* (1986). On the other hand, the infants of Putet *et al.* (1987) received a diet of pooled human milk which was likely to be relatively low in calcium and phosphorus content, in keeping with a previous study (Putet *et al.*, 1984), even if sodium supplementation (not reported) was given.

Similar effects of protein, energy and mineral supplementation have been obtained in infants fed mother's milk. When mother's own milk was supplemented with protein, fat and minerals (prepared from banked pooled human milk) rates of weight gain of 22 g·kg⁻¹d⁻¹ were obtained (Schanler *et al.*, 1985).

Longer term detailed studies on energy balance have not been carried out. However, the long term effects of different diets on weight gain of newborn infants have been studied. In a large multicentre clinical trial of infants randomly allocated to banked human milk of very low energy content (1.9 MJ/liter with 10.7 g/liter of protein) or preterm formula (3.3 MJ/liter and 20 g/liter of protein), Lucas *et al.* (1984) found significant differences in weight gain (13 vs 18 g·kg⁻¹d⁻¹), gain in length (1.0 vs 1.4 mm/day) and gain in head circumference (1.2 vs 1.6 mm/day). Tyson *et al.* (1983) found a similar effect in a comparison of banked human milk of low energy content (2.6 MJ/liter) with preterm formula (3.6 MJ/liter). They found not only large differences in weight gain (15 vs 30 g·kJ·kg⁻¹d⁻¹), gain in length (7 vs 11 mm/week) and in head circumference (8 vs 12 mm/week), but significant differences in developmental outcomes manifest in the Brazelton Neonatal Assessment Scale measurements of response to inanimate stimuli. Thus long term limitation of energy intake   has significant effects on subsequent growth and development of low birthweight infants.

6.   Dietary Pattern and Energy Intake in Infancy

Until the last fifty years, milk was the only source of food for most infants in the first year of life. Formulas have been developed from cow's milk which attempt to reproduce the macronutrient content of human milk. While the intake and composition of human milk in breast-feeding infants has been an obvious standard on which to base recommendations for the whole infant population, it has not been possible until the recent description of the nipple shield collection technique (Lucas et al., 1980) to measure the amount and composition of human milk consumed without changing the conditions of feeding, which action itself introduces errors of sampling. An alternative method uses milk collected from the breast contralateral to that used for feeding (Butte et al., 1984; Garza et al., 1986). As milk expression produces milk samples of higher energy content than does nipple-shield sampling (Williams et al., 1985), this technique may overestimate the energy intake of breast-feeding infants. This would seem to be confirmed by estimates of metabolizable energy obtained using the dual isotope technique of 2.4 MJ/liter at five weeks and 2.5 MJ/liter at eleven weeks (Lucas et al., 1987). As the preterm infant is usually gavage-fed with expressed human milk, it is appropriate to measure the composition of such milk provided it is sampled from a full 24-hr collection (Anderson et al., 1981).

There is substantial variation in the energy content of human milk. Between-mother variation is reported as 16% (Stuff et al., 1986), but there is a substantial within-mother correlation, (Hytten, 1954c; Anderson et al., 1981) which could allow mothers to be identified as high- or low-energy lactaters. Some of the other variables identified are gestational and postnatal age (Anderson, 1981): the milk of mothers delivering preterm has a slightly higher energy content (3.0 MJ/liter) than that of mothers delivering at term (2.5 MJ/liter), and these maximum values are both reached in the first two weeks of lactation. Thereafter the energy content of milk tends to remain constant until lactation ends, when it falls rather abruptly (Butte et al., 1984). Milk energy content varies with the maternal diet (Jelliffe and Jelliffe, 1978). Milk at the beginning of a feed has a lower fat and, therefore, energy content than milk collected at the end of a feed (Hytten, 1954a; Lucas et al., 1980; Williams et al., 1985). The first lactation of the day has a lower energy content than that at midday (Hytten, 1954b). The fatty acid composition of human milk varies with that of the maternal diet (Harris et al., 1984), and in turn the fatty acid composition of the milk is reflected in that of the adipose tissues of the infant. In animal models, functional consequences in brain membrane and enzyme function can be demonstrated (Clandinin, 1985).

The technique of test-weighing used to measure the milk intake of suckling infants improved with the introduction of computerized animal test-weighing scales (Butte *et al.*, 1984; Garza *et al.*, 1986). In a recently conducted survey, the daily energy intake of a large population of breastfed infants of selected well nourished mothers was 460 kJ·kg⁻¹d⁻¹at one month of age, falling to 300 kJ·kg⁻¹d⁻¹ at four months of age, values much lower than previous estimates. Furthermore, when compared to the National Growth Standards for infant growth (based largely on the growth of infants fed formula), this apparently healthy cohort of breastfed infants gained substantially less weight and had a much lower weight-for-height ratio by six months of age. The gross energy intakes described for these infants correspond well with the metabolizable energy intakes of healthy breast-fed infants determined by the dual isotope technique (400 kJ·kg⁻¹d⁻¹ at 5 weeks and 390 kJ·kg⁻¹d⁻¹ at twelve weeks of age) (Lucas *et al.*, 1987). These observations call for a substantial reexamination of current expectations for growth, and recommendations for energy intake, in the first year of life.

Relatively little is known about the energy balance of infants on mixed feeding. Solid foods were introduced as part of the normal infant diet in the early 20th century; before that time the infant diet for the first year consisted almost entirely of milk or modified milks (Cone *et al.*, 1981). The recommended age for the introduction of solid food was reduced to six months, and even earlier, in the 1950's. Very early introduction (before one month) was frequently seen throughout the 1960's and into the early 1970's. Since then, increasing speculation about the effect of early introduction of solid foods on the subsequent occurrence of obesity resulted in recommendations that this practice be delayed until five to six months of age (Fomon *et al.*, 1979), at which time the chief indication for its use is to improve iron intake. The dual isotope technique enables long term studies to be made on energy balance of infants in the first year of life. The first such study (Roberts *et al.*, 1988) compared the energy balance at three months of age of six infants born to lean mothers with that of twelve infants born to obese mothers. All infants were followed up until the age of one year. None of the lean-mother group and six of the obese-mother group of infants subsequently became obese. Despite similar metabolizable intakes (395, 398, and 367 kJ·kg⁻¹d⁻¹) the six infants who became obese had significantly lower rates of energy expenditure (256 kJ·kg⁻¹d⁻¹) than did babies who did not become obese, whether born to lean mothers (323 kJ·kg⁻¹d⁻¹) or to obese mothers (324 kJ·kg⁻¹d⁻¹). This report is consistent with the growing body of information which indicates that obesity is a predominantly genetic disorder of the control of thermogenesis (Hirsch and Leibel, 1988).

7.   Conclusion

The energy balance of infants differs substantially from that of adults because of differences in energy digestibility and in the proportion of energy intake used for growth. It is inappropriate for infant foods to be labelled with the metabolizable energy content of the food for young adults. The digestibility of infant formulas has improved greatly with the substitution of unsaturated fats of vegetable origin for butterfat, to the point that further attempts to improve digestibility with the use of medium-chain triglyceride are ineffective and may be undesirable. Structural and compositional effects of different types of formula fat and cholesterol have only recently been studied in detail and will likely be the basis for new modifications in infant formula in the future.

The energy content of human milk is not fixed, but varies with gestational and postnatal age. The practice of milk banking, particularly milk from term mothers banked to feed preterm infants, has been repeatedly shown to result in milk of low energy content, with adverse effects on infant growth and development. Further encouragement of mothers of preterm infants to express milk for their own infants hopefully will make the practice of banking obsolete. Carefully conducted clinical trials can provide guidance as to the suitability of energy and other supplementation of human milk for growing low-birthweight infants.

Energy intake and its relationship to the intake of other nutrients clearly has an impact on the amount and composition of tissue growth in preterm infants, and, as shown using even crude techniques to estimate the composition of growth, can affect the relationship between lean body mass and body weight. Despite the attention given to such studies, no long term follow up data have been reported, and the clinical significance of these short term observations remains obscure. There is a clear need to develop a technology for the accurate and non-invasive measurement of body composition of growing low-birthweight infants.

It is likely that current estimates of the energy intakes of healthy growing infants are overestimates. There is increasing evidence for a role of genetics in the regulation of energy expenditure and, consequently, in energy balance and obesity. Introduction of high energy cereal foods in the first year of life may have benefits in terms of iron balance or of ease and convenience of feeding, but there remains no justification for using these foods in terms of energy balance, growth or development. The rediscovery of the dual isotope technique and its application to long term studies of both normal and abnormal infant energy balance may provide new information on the pathogenesis of obesity and of impaired growth.

References

Abdulrazzaq, Y. M. and Brooke, O. G., 1984, Respiratory metabolism in preterm infants: the measurement of oxygen consumption during prolonged periods, *Pediatr. Res.* **18**:928.

Adam, P. A. J., 1971, Control of glucose metabolism in the human fetus and newborn infant, *Adv. Metab. Dis.* **5**:183.

Anderson, G. H., Atkinson, S. A., and Bryan M. H., 1981, Energy and macronutrient content of human milk during early lactation from mothers giving birth prematurely and at term, *Am. J. Clin. Nutr.* **34**:258.

Armsby, H. P., 1917, *The Nutrition of Farm Animals*, Macmillan, New York.

Baba, N., Bracco, E. F., and Hashim, S. A., 1982, Enhanced thermogenesis and diminished deposition of fat in response to overfeeding with diet containing medium-chain triglycerides, *Am. J. Clin. Nutr.* **35**:678.

Bach, A. C., and Babayan, V. K., 1982, Medium-chain triglycerides: an update, *Am. J. Clin. Nutr.* **36**:950.

Baldwin, R. L. and Smith, N. E., 1974, Molecular control of energy metabolism, in: *The Control of Metabolism* (J. D. Sink, ed.), pp. 17-34, Pennsylvania State University Press.

Battaglia, F. C., and Meschia, G., 1978, Principal substrates of fetal metabolism, *Physiol. Rev.* **58**:499.

Baumgart, S., Engle, W. D., Fox, W. W., and Polin, R. A., 1981, Effect of heat shielding on convection and evaporation, and radiant heat transfer in the premature infant, *J. Pediatr.* **99**:245.

Bell, E. F., and Rios, G., 1983, Air versus skin temperature servocontrol of infant incubators, *J. Pediatr.* **103**:954.

Bell, E. F., Rios, G., and Wilmoth, P. K., 1986, Estimation of 24-hour energy expenditure from shorter measurement periods in premature infants, *Pediatr. Res.* **20**:646.

Benedict, F. G., and Talbot, F. B., 1915, *The Physiology of the Newborn Infant: Character and Amount of Katabolism*, Washington, Carnegie Institute (Publication 233).

Benito, M., 1985, Contribution of brown fat to the neonatal thermogenesis, *Biol. Neonate* **48**:245.

Bhakoo, O. N., and Scopes, J. W., 1970, Minimal rates of oxygen consumption in small-for-dates babies during the first days of life, *Arch. Dis. Child.* **49**:583.

Brody, S., 1945, Methods in Animal Calorimetry, in: *Bioenergetics and Growth*, pp. 307-350, Reinhold Publishing Corporation, Waverley Press, Baltimore.

Brooke, O. G., 1980, Energy balance and metabolic rate in preterm infants fed with standard and high-energy formulas, *Br. J. Nutr.* **44**:13.

Brooke, O. G., and Alvear, J., 1982, Postprandial metabolism in infants of low birthweight, *Human Nutr. (Clin. Nutr.)* **36C**:167.

Brooke, O. G., Alvear, J., and Arnold, M., 1979, Energy retention, energy expenditure, and growth in healthy immature infants, *Pediatr. Res.* **13**:215.

Bruck, K., 1961, Temperature regulation in the newborn infant, *Biol. Neonat.* **3**:65.

Buetow, K. C., Klein, P. H., and Klein, S. W., 1964, Effect of maintenance of "normal" skin temperature on survival of infants of low birth weight. *Pediatrics* **34**:163.

Butte, N. F., Garza, C., Smith, E. O., and Nichols, B. L., 1984, Human milk intake and growth performance of exclusively breast-fed infants, *J. Pediatr.* **104**:187.

Chappell, J. E., Clandinin, M. T., Kearney-Volpe, C., Reichman, B., and Swyer P. R., 1986, Fatty acid balance studies in premature infants fed human milk of formula: effect of calcium supplementation, *J. Pediatr.* **108**:439.

Chessex, P., Reichman, B. L., Verellen, G. J. E., Putet, G., Smith, J. M., Heim, T., and Swyer, P. R., 1981, Influence of postnatal age, energy intake, and weight gain on energy metabolism in the very low-birth-weight infant, *J. Pediatr.* **99**:761.

Chessex, P., Reichman, B. L., Verellen, G. J. E., Putet, G., Smith, J. M., Heim, T., and Swyer, P. R., 1983, Quality of growth in premature infants fed their own mothers' milk, *J.Pediatr.* **102**:107.

Clandinin, M. T., Field, C. J., Hargreaves, K., Morson, L., and Zsigmond, E., Role of diet fat in subcellular structure and function, *Can. J. Physiol. Pharm.* **63**:546.

Cone, T. E., Jr., 1981, History of infant and child feeding: From the earliest years through the development of scientific concepts, in: *Infant and Child Feeding* (J. T. Bond, L. J. Filer, Jr., G. A. Leveille, A. M. Thomson, and W. B. Weil, eds.), pp. 3-34, Academic Press, New York.

Day, R., 1943, Respiratory metabolism in infancy and childhood, *Am. J. Dis. Child.* **65**:376.

Ferre, P., Decaux, D. F., Issad, T., and Girard, J., 1986, Changes in energy metabolism during the suckling and weaning period in the newborn, *Reprod. Nutr. Dev.* **26**:619.

Fomon, S. J., 1967, Body composition of the male reference infant during the first year of life, *Pediatrics* **40**:863.

Fomon, S. J, 1974, Urinary excretion of nitrogen by normal fullsize infants: relation to intake of nitrogen, in: *Infant Nutrition*, pp. 542-548, Saunders, Philadelphia.

Fomon, S. J., Ziegler, E. E., Thomas, L. M., Jensen R. L., and Filer L. J., 1970, Excretion of fat by normal term infants fed various milks and formulas, *Am. J. Clin. Nutr.* **28**:1299.

Fomon, S. J., Filer, L. J., Jr., Anderson, T. A., and Ziegler, E. E., 1979, Recommendations for feeding normal infants, *Pediatrics* **63**:52

Freymond, D., Schutz, Y., Decombaz, J. M., Micheli, J. L, and Jequier, E., 1986, Energy balance, physical activity, and thermogenic effect of feeding in premature infants, *Pediatr. Res.* **20**:638.

Garza, C., Stuff, J., and Butte, N. F., 1987, Growth of the breast-fed infant, in: *Human Lactation 3: The Effects of Human Milk Upon the Recipient Infant* (A. S. Goldman, S. A. Atkinson and L. A. Hanson, eds.), Plenum Press, New York.

Gentz, J., and Kellum, M., 1971, Metabolic effects of feeding after times of starvation, *Biol. Neonate* **19**:24.

Gentz, J., Kellum, M., and Persson, B., 1976, The effect of feeding on oxygen consumption, RQ and plasma levels of glucose, FFA, and D-β-hydroxybutyrate in newborn infants of diabetic mothers and small for gestational age infants, *Acta Paediatr. Scand.* **65**:445.

Glass L., Silverman, W. A., and Sinclair, J. C., 1968, Effect of the thermal environment on cold resistance and growth of small infants after the first week of life, *Pediatrics* **41**:1033.

Gudinchet, F., Schutz, Y., Micheli, J-L., Stettler, E., and Jequier, E., 1982, Metabolic cost of growth in very low-birth-weight infants, *Pediatr. Res.* **16**:1025.

Hamosh, M., 1979, A review: Fat digestion in the newborn: role of lingual lipase and preduodenal digestion, *Pediatr. Res.* **13**: 615.

Hamosh, M, 1982, Lingual and breast milk lipases, *Adv. Paediatr.* **29**:33

Harris, W. S., Connor, W. E., and Lindsey S., 1984, Will dietary ω-3 fatty acids change the composition of human milk? *Am. J. Clin. Nutr.* **40**:780.

Heim, T., 1981, Homeothermy and its metabolic cost, in: *Scientific Foundations of Pediatrics* (J. A. Davis and J. Dobbing, eds.), pp. 91-128, Heinemann, London.

Herrington, L. P., 1940, The heat regulation of small laboratory animals at various environmental temperatures, *Am. J. Physiol.* **129**:123.

Hey, E. N., 1969, The relation between environmental temperature and oxygen consumption in the new-born baby, *J. Physiol.* **200**:603.

Hey, E. N., 1973, Physiologic principles involved in the care of the preterm human infant, in: *The Mammalian Fetus in Vitro* (C. R. Austin, ed.), pp. 251-333.

Hey, E. N., and Katz, G., 1969, Evaporative water loss in the newborn baby, *J.Physiol.* **200**:605.

Hey, E. N. and Katz, G., 1970, The optimum thermal environment for naked babies, *Arch. Dis. Child.* **45**:328.

Hey, E. N., and Mount, L. E., 1967, Heat losses from babies in incubators, *Arch. Dis. Child.* **42**:75.

Hey, E. N., and O'Connell, B., 1970, Oxygen consumption and heat balance in the cot-nursed baby, *Arch. Dis. Child.* **45**:335.

Hill, J. R., 1959, The oxygen consumption of newborn and adult mammals: its dependance on the oxygen tension in the inspired air and on the environmental temperature, *J. Physiol.* **149**:346.

Hill, J. R., and Robinson, D. C., 1968, Oxygen consumption in normally grown, small-for-dates and large-for-dates new-born infants, *J. Physiol.* **199**:685.

Hirsch, J., and Leibel, R. L., 1988, New light on obesity, *New Eng. J. Med.* **318**:509

Hull, D., 1966, The structure and function of brown fat, *Br. Med. Bull.* **22**:92.

Hutchinson-Smith, B., 1973, Skinfold thickness in relation to birthweight, *Develop. Med. Child Neurol.* **15**:628

Hytten, F. E., 1954a, Clinical and chemical studies in human lactation: II. Variation in major constituents during a feeding, *Br. Med. J.* **1**:176.

Hytten, F. E., 1954b, Clinical and chemical studies in human lactation: IV. Individual differences in composition of milk, *Br. Med. J.* **1**:253.

Hytten, F. E., 1954c, Clinical and chemical studies in human lactation: III. Diurnal variation in major constituents of milk, *Br. Med. J.* **1**:179.

Jelliffe, D. B., and Jelliffe, E. F., 1978, The volume and composition of human milk in poorly nourished communities: a review, *Am. J. Clin. Nutr.* **31**:492.

Jones, P. J. H., Winthrop, A. L., Schoeller, D. A., Swyer, P. R., Smith, J., Filler, R. M., and Heim, T., 1987, Validation of doubly labeled water for assessing energy expenditure in infants, *Pediatr. Res.* **21**:242.

Karlberg, P., 1952, Determination of standard energy metabolism (basal metabolism) in normal infants, *Acta Pediatr. Scand.* **41** (Suppl. 89):1.

Kashyap, S., Forsyth, M., Zucker C., Ramakrishnan, R., Dell, R. B., and Heird, W. C., 1986, Effects of varying protein and energy intakes on growth and metabolic response in low birthweight infants, *J. Pediatr.* **108**:955.

Kien, C. L., Sumners, J. E., Stetina, J. S., Heimler, R., and Grausz, J. P., 1982, A method for assessing carbohydrate energy absorption and its application to premature infants, *Am. J. Clin. Nutr.* **36**:910.

Kleiber, M., 1975a, Life as a combustion process, in: *The Fire of Life; an Introduction to Animal Energetics*, pp. 3-8, Wiley, New York.

Kleiber, M., 1975b, Body size and metabolic rates, in: *The Fire of Life; an Introduction to Animal Energetics,"* pp. 179-222, Wiley, New York.

Kloosterman, G. J., 1970, On intrauterine growth: The significance of prenatal care, *Int. J. Gynaecol. Obstet.* **8**::895.

Kurzner, S. I., Garg, M., Bautista, D. B., Sargent, C. W., Bowman, M., and Keens, T. G., 1988, Growth failure in bronchopulmonary dysplasia: elevated metabolic rates and pulmonary mechanics, *J. Pediatr.* 112:73

Lemons, J. A., Moorehead, H., Jansen, R. D., and Schreiner, R. L., 1982, The energy content of infant formulas, *Early Human Dev.* 6:305.

Levine, S. Z., Wilson, J. R., Berliner, F., and Rivkin, H., 1927, The respiratory metabolism in infancy and childhood; VI. The specific dynamic action of food in normal infants, *Am. J. Dis. Child.* 35:723.

Lister, G., Hoffman, J. I. E., and Rudolph, A. M., 1974, Oxygen uptake in infants and children: a simple method for measurement, *Pediatrics* 53:656.

Lucas, A., Lucas, P. J., and Baum, D., 1980, The nipple shield sampling system: a device for measuring the dietary intake of breast-fed infants, *Early Hum. Dev.* 4:365.

Lucas, A., Gore, S. M., Cole, T. J., Bamford, M. F., Dossetor, J. F. B., Barr, I., Dicarlo, L., Cork, S., and Lucas, P. J., 1984, Multicentre trial on feeding low birthweight infants: effects of diet on early growth, *Arch. Dis. Child.* 59:722.

Lucas, A., Ewing, G., Roberts, S. B., and Coward, W. A., 1987, How much milk does the breast fed infant consume and expend? *Br. Med. J.* 29:75.

MacLean, W. C., Jr., and Fink, B. B., 1980, Lactose malabsorption by premature infants: magnitude and clinical significance, *J. Pediatr.* 97:383.

McKeown, T., and Record, R. G., 1952, Observations on fetal growth in multiple pregnancy in man, *J. Endocrinol.* 8:386.

Malin, S. W., and Baumgart, S. B., 1987, Optimal thermal management for low birth weight infants nursed under high-powered radiant warmers, *Pediatrics* 79:47.

Markestad, M. D., and Fitzharding, P. M., 1981, Growth and development in children recovering from bronchopulmonary dysplasia, *J. Pediatr.* 98:597.

Marks, K. H., Uhrman, S. B., Friedman, Z., and Misels, M. J., 1977, The effect of clothing on the growth of very low birth weight infants, *Pediatr. Res.* 11:540.

Merrill, A. L. and Watt, B. K., 1973, Energy value of foods: basis and derivation. Agriculture Handbook No. 74, United States Department of Agriculture, Washington.

Mestyan, J., Jarai, I., and Fekete, M., 1968, The total energy expenditure and its components in premature babies maintained under different nursery and environmental conditions, *Pediatr. Res.* 2:161.

Mestyan, J., Jarai, I., Kekete, M., and Soltesz, G., 1969, Specific dynamic action in premature infants kept at and below the neutral temperature, *Pediatr. Res.* 3:41.

Moss, M., Moreau G., and Lister, G., 1987, Oxygen transport and metabolism in the concious lamb: the effects of hypoxia, *Pediatr. Res.* 22:177.

Murlin, J. R., Conklin, R. E., and Marsh, M. E., 1925, Energy metabolism of normal new-born babies, *Am. J. Dis. Child.* 29:1.

Okamoto, E., Muttart, C. R., Zucker, C. L., and Heird, W. C., Use of medium-chain triglycerides in feeding the low-birth-weight infant, *Am. J. Dis. Child.* 136:428.

Perlstein, P. H., Edwards, N. K., Atherton, H. D., and Sutherland, S. M., 1976, Computer-assisted newborn intensive care, *Pediatrics* 57:494.

Persson, B., and Gentz, J., 1966, The pattern of blood lipids, glycerol and ketone bodies during the neonatal period, infancy and childhood, *Acta. Paediatr. Scand.* 55:353

Persson, P-H., and Weldner, B-M., 1986, Intrauterine weight curves obtained by ultrasound, *Acta Obstet. Gynecol. Scand.* 65:169.

Pullar, J. D., and Webster, A. J. F., 1977, The energy cost of fat and protein deposition in the rat, *Br. J. Nutr.* 37:355

Putet, G., Senterre, J., Rigo, J., and Salle, B., 1984, Nutrient balance, energy utilization, and composition of weight gain in very-low-birth-weight infants fed pooled human milk or a preterm formula, *J. Pediatr.* 105:79.

Putet, G., Rigo, J., Salle, B., and Senterre, J., 1987, Supplementation of pooled human milk with casein hydrolysate: energy and nitrogen balance and weight gain composition in very low birth weight infants, *Pediatr. Res.* 21:458.

Reichman, B. L., Chessex, P., Putet, G., Verellen, G. J. E., Smith, J. M., Heim, T., and Swyer, P. R., 1981, Diet, fat accretion, and growth in premature infants, *N. Eng. J. Med.* 305:1495.

Reichman, B. L., Chessex, P., Putet, G., Verellen, G. J. E., Smith, J. M., Heim, T., and Swyer, P. R., 1982, Partition of energy metabolism and energy cost of growth in the very-low-birth-weight infant, *Pediatr.* 69:446.

Reichman, B. L., Chessex, P., Verellen, G. J. E., Putet, G., Smith, J. M., Heim, T., and Swyer, P. R., 1983, Dietary composition and macronutrient storage in preterm infants, *Pediatr.* 72:322.

Richardson, P., Bose, C. L., Bucchiarelli, R. L., and Carlstrom, J. R., 1984, Oxygen consumption of infants with respiratory distress syndrome, *Biol. Neonat.* 46:53.

Roberts, S. B., Savage, J., Coward, W. A., Chew, B., and Lucas, A., 1988, Energy expenditure and intake in infants born to lean and overweight mothers, *N. Eng. J. Med.* 318:461

Romney, S. L., Reid, D. E., Metcalfe, J., and Burwell, C. S., 1955, Oxygen utilization by the human fetus *in utero, Am. J. Obstet. Gynecol.* 70:791.

Rubecz, I., and Mestyan, J., 1975, The partition of maintenance energy expenditure and the pattern of substrate utilization in intrauterine malnourished newborn infants before and after recovery, *Acta Pediatr. Hung.* 16:335.

Sauer, P. J. J., Dane, H. J., and Visser, H. K. A., 1984, Longitudinal studies of metabolic rate, heat loss, and energy cost of growth in low birthweight infants, *Pediatr. Res.* 18:254.

Schanler, R., Garza C., and Nichols, B. L., 1985, Fortified mothers' milk for very low birth weight infants: results of growth and nutrient balance studies, *J. Pediatr.* 107:437.

Schulte, F. J., 1981, Developmental Neurophysiology, in: *Scientific Foundations of Pediatrics* (J. A. Davis and J. Dobbing, eds.), pp. 785-829, Heinemann, London.

Schulze, K. F., Stefanski, M., Masterton, J., Spinnazola, R., Ramakrishnan, R., Dell, R. B., and Heird, W. C., 1987, Energy expenditure, energy balance and composition of weight gain in low birth weight infants fed diets of different protein and energy content, *J. Pediatr.* 110:753.

Scopes, J. W., and Ahmed, I., 1966, Indirect assessment of oxygen requirements in newborn babies by monitoring deep body temperature, *Arch. Dis. Child.* 41:25.

Senterre, J., and Karlberg, P., 1970, Respiratory quotient and metabolic rate in normal full-term and small-for-date newborn infants, *Acta Pediat. Scand.* 59:653.

Silver, M., 1976, Fetal Energy Metabolism, in: *Fetal Physiology and Medicine* (R. W. Beard and P. W. Nathanielsz, eds.), pp. 173-193, Saunders, Philadelphia.

Silverman, W. A., Fertig, J. W., and Berger, A. P., 1958, The influence of the thermal environment upon the survival of newly born premature infants, *Pediatrics* 22:876.

Sinclair, J. C., 1976, Metabolic rate and temperature control, in: *The Physiology of the Newborn Infant,* 4th ed. (C. A. Smith and N. M. Nelson, eds.), pp. 354-415, Thomas, Springfield, IL.

Sinclair, J. C., 1978, The Energy Balance of the Newborn, in: *"Temperature Regulation and Energy Metabolism in the Newborn* (J. C. Sinclair, ed.), pp. 187-204, Grune and Stratton, New York.

Sinclair, J. C., and Silverman, W. A., 1966, Intrauterine growth in active tissue mass of the human fetus, with particular reference to the undergrown baby, *Pediatr.* **38**:48.

Sinclair, J. C., Scopes, J. W., and Silverman, W. A., 1967, Metabolic reference standards for the neonate, *Pediatr.* **39**:724.

Sparks, J. W., Girard, J. R., and Battaglia, F. C., 1980, An estimate of caloric requirements of the human fetus, *Biol. Neonate* **38**:113.

Stothers, J. K., 1981, Head insulation and heat loss in the newborn, *Arch. Dis. Child,* **56**:530.

Stothers, J. K., and Warner, R. M., 1979, Effect of feeding on neonatal oxygen consumption, *Arch. Dis. Child.* **54**:415.

Stothers, J. K., and Warner, R. M., 1984, Thermal balance and sleep state in the newborn, *Early Hum. Dev.* **9**:313.

Stuff, J., Garza, C., Fraley, J. K., Smith, E. O., Klein, E. R., and Nichols, B. L., 1986, Sources of variation in milk and caloric intakes in breast-fed infants: implications for lactation study design and interpretation, *Am. J. Clin. Nutr.* **43**:361.

Swyer, P. R., 1978, Heat loss after birth, in: *Temperature Regulation and Energy Metabolism in the Newborn* (J. C. Sinclair, ed.), pp.91-128, Grune and Stratton, New York.

Tanner, J. M., 1970, Standards for birthweight or intrauterine growth, *Pediatrics* **46**:1.

Tantibhedhyangkul, P., and Hashim, S. A., 1975, Medium chain triglyceride feeding in premature infants: effects on fat and nitrogen absorption, *Pediatrics* **55**:359.

Tyson, J. E., Lasky, R. E., Mize, C. E., Richards, C. J., Blair-Smith, N., Whyte, R., and Beer, A. E., 1983, Growth, metabolic response, and development in very-low-birth-weight infants fed banked human milk or enriched formula: I. Neonatal findings, *J. Pediatr.* **103**:95.

Usher, R., and MacLean, F., 1969, Intrauterine growth of live-born Caucasian infants at sea level: standards obtained by measurements in 7 dimensions of infants born between 25 and 44 weeks gestation, *J.Pediatr.* **74**:901.

Villee, C. A., 1953, The metabolism of human placenta *in vitro*, *J. Biol. Chem.* **205**:113.

Weinstein, M. R., and Oh, W., 1981, Oxygen consumption in infants with bronchopulmonary dysplasia, *J.Pediatr.* **99**:958.

Weinstein, M. R., Haugen, K., Bauer, J. H., Hewitt, J., and Finan, D., 1987, Intravenous energy and amino acids in the preterm newborn infant: Effects on metabolic rate and potential mechanisms of action, *J. Pediatr.* **111**: 119.

Whyte, R. K., Sinclair, J. C., Bayley, H. S., Campbell, D., and Singer, J., 1982, Energy cost of growth of premature infants, *Acta Pediatr. Acad. Sci. Hung.* **23**:85.

Whyte, R. K., Haslam, R., Vlainic, C., Shannon, S., Samulski, K., Campbell, D., Bayley, H. S., and Sinclair, J. C., 1983, Energy balance and nitrogen balance in growing low birthweight infants fed human milk or formula, *Pediatr. Res.* **17**:891.

Whyte, R. K., Bayley, H. S., and Sinclair, J. C., 1985, Energy intake and the nature of growth in low birth weight infants, *Can. J. Physiol. Pharmacol.* **63**:565.

Whyte, R. K., Campbell D., Stanhope, R., Bayley, H. S., and Sinclair, J. C., 1986, Energy balance in low birth weight infants fed formula of high or low medium-chain triglyceride content, *J. Pediatr.* **108**:964

Widdowson, E. M., and Dickerson, J. W. T, 1964, Chemical composition of the body, in: *Mineral Metabolism: an Advanced Treatise* (C. L. Comar and F. Bronner, eds.), pp. 1-247, Academic Press, New York.

Ziegler, E. E., O'Donnell, A. M., Nelson, S. E., and Fomon, S. J., 1976, Body composition of the reference fetus, *Growth* **40**:329.

Chapter 5

Nutritional Assessment of the Hospitalized Patient

Paul M. Starker

1. Introduction

Nutritional assessment is a technique by which malnutrition can be identified. The importance of identifying malnourished patients has greatly increased in the recent past because of our increased ability to alter the nutritional state. Significant malnutrition remains a term which is difficult to define because it implies that nutritional intervention is mandated to improve clinical outcome.

The frequency of malnutrition in hospitalized patients has been estimated to be as high as 50% (Bistrian et al., 1974; Bistrian et al., 1976; Kassiadou et al., 1978). The implications of this frequency in terms of clinical outcome are staggering. The purpose of this discussion is to identify the common abnormalities associated with malnutrition and to review the methods by which clinicians attempt to identify these abnormalities accurately. Finally, the clinical implications of malnutrition in the hospitalized patient and the results of nutritional intervention will be discussed.

Paul M. Starker • Department of Surgery, Columbia-Presbyterian Medical Center, New York, NY 20032.
Advances in Nutritional Research, Vol. 8
Edited by Harold H. Draper
Plenum Press, New York, 1990

2.   Malnutrition

Malnutrition may be viewed as an alteration in dietary intake that results in changes in subcellular, cellular and/or organ function which expose the patient to increased risk of morbidity and mortality (Grant, 1986). These complications generally can be related to the alterations in body composition and immune function that accompany progressive malnutrition.

Changes in body composition seen during malnutrition classically involve redistribution of total body water, which normally constitutes 50-70% of total body weight. In health about 40% of total body water is found in the extracellular space and about 60% in the intracellular space. As malnutrition progresses the percentage of total body water found in the extracellular space increases. Elwyn *et al.* (1975) used radioisotopes to demonstrate this phenomenon. They showed that with progressive nutritional depletion the ratio of extracellular fluid to total body water rose from 0.43 to 0.59. This trend was reversed by nutritional repletion. Based on the fact that sodium is the major extracellular cation and potassium is the major intracellular cation, Shizgal *et al.* (1976) used the ratio of exchangeable sodium (Na_e) to exchangeable potassium (K_e) to follow the redistribution of total body water during nutritional depletion. They found that as the ratio of extracellular water to total body water rose from 0.40 to 0.52 the Na_e/K_e ratio rose from 0.98 to 2.07. This trend was reversed with nutritional repletion.

These methods for documenting the changes in body composition seen with malnutrition are quite sophisticated. A simpler, although less accurate, way of following these changes is by monitoring sodium balance and serum sodium. Starker *et al.* (1982) showed that with progressive nutritional depletion sodium balance was markedly positive and that with repletion this trend was reversed, a finding totally consistent with the findings of previous investigators.

The changes in immune function that accompany malnutrition are often equally as difficult to quantify. Cell-mediated immunity is markedly depressed in malnourished patients. This is evidenced by a decrease in total lymphocyte and T-lymphocyte counts, as well as a decrease in the lymphocyte response to mitogenic stimulation. Anergy to skin test reagents also is present (Law *et al.*, 1973; Haffejee and Angorn, 1979; Smythe *et al.*, 1971; Chandra, 1974).

This alteration in immune function also has been shown to correlate well with the aforementioned changes in body composition. Spanier *et al.* (1976) showed that as the Na_e/K_e ratio reached values that were consistent with malnutrition, a significant change in the patient's ability to respond to skin test reagents developed. The anergic state was reversed as body composition normalized during nutritional repletion.

The changes in body composition and immune function that accompany progressive malnutrition can be related to the increased morbidity and mortality seen in hospitalized patients suffering from malnutrition. It is the clinician's task to identify malnourished patients before complications arise so that nutritional therapy can be instituted. A multitude of tests have been developed in attempts to accurately assess the nutritional state. These will now be reviewed.

3.   Parameters of Nutritional Assessment

3.1   Anthropometry

Anthropometric measurements are used to assess energy and protein reserves. Fat has the highest caloric value. During times of energy deficit fat stores are mobilized. Measurements of fat stores, therefore, can yield information regarding the length and severity of the energy deficit. The most commonly used measurements are triceps and subscapular skinfold thickness. The sum of these two measurements gives fairly accurate and reproducible results which correlate well with changes in body weight (Bray et al., 1978; Bradfield et al., 1979).

3.2   Muscle Mass

During energy deficit the body attempts to conserve skeletal muscle mass. Although this body component can be utilized as an energy source, its use is counterproductive because of the functional importance of muscle proteins. One method of estimating protein mass is by measuring arm circumference and arm muscle area. Like skinfold thickness, measurements are compared to tables of normal values and patients are placed in percentiles. Patients who fall below the twenty-fifth percentile are judged to be severely malnourished. Unfortunately, both these anthropometric methods suffer from great variability secondary to variations in technique on the part of the observer and to clinical variables, and therefore their clinical value in an individual is often suspect (Hull et al., 1980; Collins et al., 1979).

3.3   Body Weight

A more common anthropometric measurement used to assess nutritional status is body weight. Weight as a measure of nutritional status may be expressed in a variety of ways. These include weight for height, relative body weight (observed body weight/standard reference weight x 100), percent of usual weight (actual weight/usual weight x 100), percent weight change and absolute weight loss (Seltzer *et al.*, 1982). However, recent loss of weight over a relatively short period of time is generally felt to be the most significant parameter. Percentage weight change is derived from the following formula (Grant *et al.*, 1981):

$$\frac{\text{Usual weight - Current weight}}{\text{Usual weight}} \times 100$$

Weight loss of greater than ten percent is usually deemed significant (Blackburn *et al.*, 1977).

Many authors have related weight loss to morbidity and mortality. In fact, Studley (1936) showed a relationship between weight loss and postoperative morbidity and mortality in surgery for benign peptic ulcer disease. Other authors have found similar correlations. Seltzer *et al.* (1982) found that an absolute weight loss of greater than ten pounds in the six months prior to undergoing elective surgery increased the risk of postoperative mortality at least tenfold. DeWys (1980) demonstrated that the median survival of chemotherapy patients who had sustained weight loss prior to therapy was one-half that of those who had not lost weight.

Unfortunately, the degree of weight loss often does not reflect true nutritional status. Total body weight is influenced by many factors. One of the largest contributors to body weight is total body water. During periods of malnutrition the extracellular fluid space expands, and if body weight alone is used as a nutritional parameter this phenomenon may mask the fact that protein stores are being depleted (Elwyn *et al.*, 1975; Starker *et al.*, 1982; Spanier and Shizgal, 1977). More direct measurements of protein stores are often attempted using biochemical indices.

3.4   Protein Mass

Methods for assessing somatic protein mass include 24-hr urinary creatinine excretion and creatinine/height index. Difficulties in sample collection and variations in body build make the application of these two measurements in clinical practice unrewarding. Collection of urine for 24-hr nitrogen or 3-methyl histidine excretion suffers from the same technical problems and therefore also has limited application (Grant *et al.*, 1981; McLaren and Meguio, 1983). Assessment of somatic protein stores is obviously quite difficult. Even if done accurately, size and function may not correlate. Direct measurement of skeletal muscle function can be accomplished by using hand grip dynamometry. This method is highly touted by some investigators as a form of nutritional assessment (Klidjian *et al.*, 1980, 1982) but because there are multiple uncontrolled variables and because forearm function may not represent skeletal muscle function elsewhere in the body, this method is also far from foolproof (McLaren and Meguio, 1983; Lopes *et al.*, 1982).

Visceral protein mass is usually assessed by measuring the serum concentrations of the transport proteins synthesized by the liver. Serum albumin classically has been used as a method of assessing malnutrition. Because of its long half-life (20 days) and large volume of distribution, serum albumin levels fall and recover slowly with changes in nutrition. Acute changes in serum albumin generally reflect changes in the volume of body water in which it is distributed (Starker *et al.*, 1982, 1983). However, changes in serum albumin concentration over long periods of time can reflect malnutrition of various severities (Table I).

3.5   Serum Proteins

Serum transferrin may be a more accurate reflection of an acute change in nutritional status because of its shorter half-life (8.8 days) and its smaller volume of distribution (Awai and Brown, 1963). Serum transferrin is best measured by a sophisticated biochemical assay, but can also be calculated from a formula based on total iron binding capacity (Blackburn *et al.*, 1977; Goodwin *et al.*, 1966). This method suffers from variations in iron status.

Prealbumin and retinol-binding proteins are two other liver-dependent proteins used in nutritional assessment. Prealbumin has a short half-life (2 days) and a small body pool (Oppenheimer *et al.*, 1965). Its levels are therefore quite sensitive to acute changes in protein synthesis. Retinol-binding protein has a half-life of only 10 hr (Peterson, 1971). Because of this rapid turnover it is so sensitive to changes in protein synthesis that it

Table I. Serum Albumin Concentration as a Marker of Malnutrition

Concentration (g/dl)	Level of Malnutrition
>3.5	None
2.8 - 3.5	Mild
2.1 - 2.7	Moderate
<2.1	Severe

is, unfortunately, of little use in the routine clinical setting (Grant *et al.*, 1981).

Changes in the serum concentration of these visceral proteins is a multifactorial phenomenon. One of the factors is malnutrition. Decreases in the serum level of these proteins secondary to malnutrition are associated with many other metabolic changes; hence it is likely that these decreases also reflect other metabolic disturbances.

3.6   Immunocompetence

One body function which suffers during malnutrition is immunocompetence. Immune function can be studied in a quantitative as well as a qualitative manner. Quantitatively, total lymphocyte count, T-lymphocyte count, immunoglobulin levels and complement levels are useful measurements. Cell-mediated immunity can be assessed qualitatively by a mitogen stimulation test or by delayed cutaneous hypersensitivity. Common antigens for the latter test include mumps, candida, streptokinase-streptodornase, trichophyton and tuberculin (Miller, 1978).

Progressive malnutrition has been associated with a progressive decline in immunocompetence. Lymphocyte response to mitogenic stimulation is impaired, there is a depression in skin test reactivity and the total number of lymphocytes declines (Law *et al.*, 1973; Haffejee and Angorn, 1979).

As with many other nutritional parameters, immunocompetence is a multifactorial entity. Malnutrition may cause a depression in immunocompetence; however, this depression is also seen in cancer patients and in patients on immunosuppressive drugs such as chemotherapeutic agents and steroids. Therefore, immunocompetence may have nothing to do with the nutritional state or it may be directly related to the nutritional state. Immune testing therefore should be considered as another component of the nutritional status.

4.   Prediction of Malnutrition

Since it appears that there is no one easy and accurate technique of nutritional assessment, many authors have incorporated several parameters into various prediction equations. Patients then are stratified in terms of outcome to develop a predictive model for nutritionally related complications.

Several investigators have used abnormal values for a single parameter as an indicator of malnutrition and correlated this with increased complications (Seltzer *et al.*, 1982; Bistrian *et al.*, 1974; Reinhardt *et al.*, 1980; Rhoads and Alexander, 1955). Other investigators have combined parameters in an attempt to refine the assessment techniques. Seltzer (1979) has shown that in patients with abnormalities in serum albumin and total lymphocyte count a significant increase in morbidity and mortality can be anticipated. Hickman (1980) used abnormalities in serum albumin and body weight to predict an increased rate of complications and death after colon resection.

A more sophisticated method of predicting post-operative complications was derived by Buzby et al. (1980). The Prognostic Nutritional Index (PNI) stratifies patients into categories of low, intermediate or high risk based on the following formula:

$$PNI \ (\%) = 158 - 16.6(ALB) - 0.78(TSF) - 0.2(TFN) - 5.8(DH)$$

where: ALB = Serum albumin;   TSF = Triceps skinfold;   TFN = Serum transferrin;   DH = Cutaneous delayed hypersensitivity reactivity.

Patients in the high risk group had a morbidity and mortality rate of 46% and 33%, respectively, compared to the low risk group who had rates of 8% and 3%, respectively.

5. Nutritional Intervention

The ability to predict poor outcome for a patient leads to attempts to prevent this outcome using nutritional intervention. Nutritional support has been shown to reverse abnormalities in both body composition and immune function. Body composition, measured either as the Na_e/K_e ratio (Forse et al., 1981) or as sodium balance (Starker et al., 1982) normalizes with nutritional repletion except in patients who are stressed and critically ill (Starker et al., 1982; Shizgal et al., 1976). Immune function also improves during nutritional repletion in patients who are not stressed.

An increase in total lymphocyte and T-lymphocyte counts, improvement in T and B cell function, and reversal of anergy have all been shown to occur with nutritional repletion (Law et al., 1973; Haffejee and Angorn, 1979; Spanier et al., 1976). In fact, changes in body composition and immune function have been shown to be correlated (Spanier et al., 1976; Forse et al., 1981).

Initial attempts at improving operative morbidity and mortality in malnourished patients focused on early postoperative support. This technique did not lead to any significant improvement (Abel et al., 1976; Holter and Fischer, 1977). Other investigators have shown that, with at least one week of preoperative nutritional support, morbidity and mortality in severely malnourished patients can be significantly reduced (Mullen et al., 1980; Starker et al., 1986).

The fact that malnourished patients are at a greater risk for complications than are normal patients is well documented. The ability to accurately identify those patients who are significantly malnourished has not yet been achieved. There are many parameters which can be used to assess the nutritional state but none alone is reliable. Combinations of various parameters seem to improve the ability to perform accurate nutritional assessment but the easily measured parameters are not always reliable. Sophisticated methodology may be necessary to perform accurate nutritional assessment, but that obviously is not widely available.

With the techniques for altering the nutritional state that are now available, accurate nutritional assessment becomes more important. The clinician has to be able to determine which patient will benefit from nutritional support and which patient should not be exposed to the risk of artificial supplementation routes because of the small benefits to be gained. A means of performing nutritional assessment which is accurate and simple is required to identify patients who require and will benefit from nutritional intervention.

The methods of nutritional assessment available today are simple but as yet are not as accurate as one would desire. If one had to choose any single parameter, weight loss would be the parameter chosen. Rapid, non-purposeful weight loss is the single best predictor of malnutrition we have today. When greater than 10% of body weight is lost, protein stores become depleted to the point of compromising bodily functions, especially immune function and healing capacity. Morbidity and mortality rates parallel this weight loss.

Until the day when a simple and accurate index of nutritional status is developed, change in body weight will remain an essential criterion. Depending on the institution, this criterion may be combined with various other methods of assessment to detect malnutrition in the hospitalized patient, and to indicate a need for nutritional intervention for the prevention of morbidity and mortality.

References

Abel, R. M., Fischer, J. E., Buckley, M. S., Barnett, G. O., and Austen, G., 1976, Malnutrition in cardiac surgical patients: Results of prospective randomized evaluation of early postoperative parenteral nutrition, *Arch. Surg.* 111:45.

Awai, M., and Brown, E. B., 1963, Studies of the metabolism of I-131 labelled human transferrin, *J. Lab. Clin. Med.* 61:363.

Bistrian, B. R., Blackburn. G. L., Hallowell, E., and Heddle, R., 1974, Protein status of general surgical patients, *JAMA* 230:858.

Bistrian, B. R., Blackburn, G. L., Vitale, J., Cochran, D., and Naylor, J., 1976, Prevalence of malnutrition in general medical patients, *JAMA* **235**:1567.

Blackburn, G. L., Bistrian, B. R., Maini, B. S., Schlamm, H. T., and Smith, M. F., 1977, Nutritional and metabolic assessment of the hospitalized patient, *J. Parent. Ent. Nutr.* **1**:11.

Bradfield, R. B., Schutz, Y., and Lechtig, A., 1974, Skinfold changes with weight loss, *Am. J. Clin. Nutr.* **32**:175.

Bray, G. A., Greenway, F. L., Molitch, M. E., Dahms, W. T., Atkinson, R. L., and Hamilton, K., 1978, Use of anthropometric measurements to assess weight loss, *Am. J. Clin. Nutr.* **31**:769.

Buzby, G. P., Mullern, J. L., Matthews, D. C., Hobbs, C. L., and Rosato, E. F., 1980, Prognostic nutritional index in gastrointestinal surgery, *Am. J. Surg.* **139**:160.

Chandra, R. K., 1974, Rosette-forming T-lymphocytes and cell-mediated immunity in malnutrition, *Br. Med. J.* **3**:608.

Collins, J. P., McCarthy, I. D., and Hill, G. L., 1979, Assessment of protein nutrition in surgical patients--the value of anthropometrics, *Am. J. Nutr.* **32**:1527.

Dewys, D. M., 1980, Nutritional care of the cancer patient, *JAMA* **224**:374.

Elwyn, D. H., Bryan-Brown, C. W., and Shoemaker, W. C., 1975, Nutritional aspects of body water dislocations in postoperative and depleted patients, *Ann. Surg.* **182**:76.

Forse, R. A., Christou, N. U., Meakins, J. L., MacLean, L. D., and Shizgal, H. M., 1981, The reliability of skin testing as a measure of the nutritional state, *Arch. Surg.* **116**:1284.

Goodwin, J. F., Murphy, B., and Guillernette, M., 1966, Direct measurement of serum iron and binding capacity, *Clin. Chem.* **12**:247.

Grant, J. P., 1986, Nutritional assessment in clinical practice, *Nutr. Clin. Practice*, **1**:3.

Grant, J. P., Custer, P. B., and Thurlow, J., 1981, Current techniques of nutritional assessment, *Surg. Clin. N. Amer.* **61**:437.

Haffejee, A. A., and Angorn, I. B., 1979, Nutritional status and the nonspecific cellular and humoral immune response in esophageal carcinoma, *Ann. Surg.* **189**:475.

Hickman, D. M., Miller, R. A., Rombean, J. L., Twomey, P. L., and Frey, C. F., 1980, Serum albumin and body weight as predictors of postoperative course in colorectal cancer, *J. Parent. Ent. Nutr.* **4**:314.

Holter, A. R., and Fischer, J. E., 1977, The effects of perioperative hyperalimentation on complications in patients with carcinoma and weight loss, *J. Surg. Res.* **23**:31.

Hull, J. C., O'Quigley, J., Giles, G. R., Appleton, N., and Stocks, H., 1980, Upper limb anthropometry: The value of measurement variance studies, *Am. J. Clin. Nutr.* **33**:1846.

Kassiadou, A., Domingos, J. C., Vianna, R., and Clinc, G., 1978, Malnutrition in general medical and surgical patients, *J. Parent. Ent. Nutr.* **2**:199 (Abst.).

Klidjian, A. M., Foster, K. J., Kammerling, R. M., Looper, A., and Kanan, S. J., 1980, Relationship of anthropometric and dynamometric variables to serious post-operative complications, *Br. J. Med.* **281**:899.

Klidjian, A. M., Archer, T. J., Foster, K. J., and Kanan, S. J., 1982, Detection of dangerous malnutrition, *J. Parent. Ent. Nutr.* **6**:119

Law, D. K., Dudrick, S. J., and Abdou, N. I., 1973, Immunocompetence of patients with protein-calorie malnutrition, *Annals Int. Med.* **79**:545.

Lopes, J. M., Russell, D. M., Whitwell, J., and Jeejeebhoy, K. N., 1982, Skeletal muscle function in malnutrition, *Am. J. Clin. Nutr.* **36**:602.

Mancini, G., Carbonara, A. O., amd Heremans, J. F., 1965, Immunological quantification of antigens by single radial immunodiffusion, *Int. J. Immunochem.* **2**:235.

McLaren, D. S., and Meguio, M. M., 1983, Nutritional assessment at the crossroads, *J. Parent. Ent. Nutr.* **7**:575.

Miller, C. L., 1978, Immunological assays as measurements of nutritional status: A review, *J. Parent. Ent. Nutr.* **2**:554.

Muller, J. L., Buzby, G. P., Matthews, D. C., Smale, B. F., and Rosato, E. F. 1980, Reduction in operative morbidity and mortality by combined preoperative and postoperative nutritional support, *Ann. Surg.* **192**:604.

Oppenheimer, J. H., Surks, M. I., Bernstein, G., and Smith, J. C., 1965, Metabolism of I-131 labelled thyroxine-binding prealbumin in man, *Science* **149**:748.

Peterson, P. A., 1971, Demonstration in serum of two physiological forms of human retinol-binding protein, *Eur. J. Clin. Invest.* **1**:437.

Reinhart, G. F., Myscofski, J. W., and Wilkens, D. B., 1980, Incidence and mortality of hypoalbuminemic patients in hospitalized veterans, *J. Parent. Ent. Nutr.* **4**:357.

Rhoads, J. E., and Alexander, C. E., 1955, Nutritional problems of surgical patients, *Ann. N. Y. Acad. Sci.* **63**:268.

Seltzer, M. H., Bastidas, J. A., Cooper, D. M., Engler, P., Slocum, B., and Fletcher, H. S., 1974, Instant nutritional assessment, *J. Parent. Ent. Nutr.* **3**:157.

Seltzer, M. H., Slocum, B. A., Cutalde-Beteher, E. L., Fileti, C., and Gerson, N., 1982, Instant nutritional assessment: Absolute weight loss and surgical mortality, *J. Parent. Ent. Nutr.* **6**:218.

Shizgal, H. M., Spanier, A. H., and Kurtz, R. S., 1976, Effect of parenteral nutrition on body composition in the critically ill patient, *Am. J. Surg.* **131**:156.

Smythe, P. M., Schonland, M., Brereton-Stiles, G. G., Mafoyane, A., Grace, H. J., Coovadia, H. M., Loening, W. E. K., Parent, M. A., and Vos, G. H., 1971, Thymolymphatic deficiency and depression of cell mediated immunity in protein-calorie malnutrition, *Lancet* **2**:939.

Spanier, A. H., and Shizgal, H. M., 1977, Caloric requirements of the critically ill patient receiving intravenous hyperalimentation, *Am. J. Surg.* **133**:99.

Spanier, A. H., Menkins, J. L., MacLean, L. D., and Shizgal, H. M., 1976, The relationship between immune competence and nutrition, *Surg. Forum* **27**:332.

Starker, P. M., Gump, F. E., Askanazi, J., Elwyn, D. H., and Kinney, J. M., 1982, Serum albumin as an index of nutritional support, *Surgery* **91**:194.

Starker, P. M., LaSala, P. A., Askanazi, J., Gump, F. E., Forse, R. A., and Kinney, J. M., 1983, The response to TPN, *Ann. Surg.* **198**:720.

Starker, P. M., LaSala, P. A., Askanazi, J., Todd, G., Hensle, T. W., and Kinney, J. M., 1986, The influence of preoperative total parenteral nutrition upon morbidity and mortality, *Surg. Gynecol. Obstet.* **162**:569.

Studley, H. O., 1936, Percentage of weight loss: A basic indicator of surgical risk in patients with chronic peptic ulcer, *JAMA* **106**:458.

Chapter 6

Nutritional Modulation of Oxygen Radical Pathology

Harold H. Draper

1.   Introduction

In the past decade the pathogenicity of oxygen radicals has emerged as a major focus of research in fundamental biology. This emergence is reflected in the appearance of numerous books and conference proceedings on the subject, of which only five are cited here (Simic *et al.*, 1988; Chow, 1988; Halliwell, 1988a; Quintanilha, 1988a; Halliwell and Gutteridge, 1985), as well as several new periodicals devoted to the publication of ongoing research. The field has been remarkable for the rapidity with which basic research findings have been incorporated into clinical medicine. New perceptions of the role of oxygen radicals in physiological processes have led to novel strategies for the prevention or treatment of a range of pathologies, including inflammmation, immunosuppression, cataracts, radiation injury, hyperoxia, cirrhosis and cancer. The prominent involvement of nutrients in the metabolism of oxygen radicals has raised the possibility that these conditions may be subject to nutritional modulation.

Harold H. Draper • Department of Nutritional Sciences, College of Biological Science. University of Guelph, Guelph, ON, N1G 2W1.

Advances in Nutritional Research, Vol. 8
Edited by Harold H. Draper
Plenum Press, New York, 1990

2.   Traditional Concept of Oxygen Toxicity

The classical concept of oxidative tissue damage envisioned a central role of lipoxy radicals, formed by a metal-catalyzed attack of molecular oxygen on the polyunsaturated fatty acids (PUFA) present in the phospholipids (PL) of cell membranes. Fatty acyl peroxy radicals propagate chain reactions among cellular lipids, causing changes in membrane structure and function, and inflicting pervasive biochemical damage on proteins, nucleic acids and other compounds. Some of their effects are readily recognizable clinically, as in the case of the damage to vascular membranes seen in a deficiency of the lipid antioxidant vitamin E, whereas others may be subclinical and still others may undergo repair.

An enzymatic component was added to this chemical concept of lipid peroxidation with the discovery of selenium-dependent glutathione peroxidase (Se-GPx), which was found to be capable of reducing fatty acyl hydroperoxides to their corresponding hydroxy acids. This discovery provided an explanation for the mysterious "sparing effect" of Se on the requirement for vitamin E, as well as an apparent mechanism for the removal of lipid hydroperoxides generated by donation of hydrogen to lipoperoxy radicals by the antioxidant action of this vitamin. For two decades thereafter, the manifestations of oxidative damage caused by such conditions as hyperoxia, iron and carbon tetrachloride poisoning, and exposure to the respiratory oxidants ozone and nitrogen oxides, were attributed to the peroxidation of lipids in an environment containing excess oxygen or metal catalysts, inadequate Se or vitamin E, or compounds that generate carbon centered radicals subject to oxygen uptake. This perception was reinforced by the capacity of the "antioxidant nutrients," vitamin E and selenium, to prevent or reduce the toxicity of hyperoxia, respiratory oxidants and some xenobiotics.

3.   Modern Concept of Oxygen Toxicity

This circumscribed view of oxygen toxicity was vastly expanded by the identification of additional so-called "antioxidant enzymes": two superoxide dismutases (SOD) that catalyze dismutation of the reactive oxygen species superoxide (O_2^-), and a non-Se-dependent glutathione peroxidase (GPx) with greater affinity than the Se-containing peroxidase for lipid hydroperoxides. Of comparable importance was the demonstration that H_2O_2 plays a major role in oxygen radical pathology as a source of highly destructive hydroxyl radicals (HO·) formed by its reaction with metal ions. This discovery also led to a recognition of the critical antioxidant roles of catalase and Se-GPx in decomposing H_2O_2 to H_2O and molecular oxygen. Hydroxyl radicals are capable of attacking most biological compounds and have been implicated in oxidative damage to numerous cellular macromolecules. They are also centrally involved in the initiation of lipid peroxidation. The enzymes involved in removing O_2^- and H_2O_2, thereby preventing HO· generation, consequently have come to be regarded as comprising the main intracellular antioxidant defense system.

There is little SOD, GPx or catalase in the blood plasma and extracellular fluids, where the chief components of the antioxidant system appear to be compounds that bind transition metal catalysts of oxygen radical generation. These compounds include transferrin, lactoferrin, ceruloplasmin, albumin and other metal-binding substances. The importance of metal binding is manifested in the erythrocyte and other cellular membrane damage caused by intravenous administration of trace quantities of ferrous ions. The failure of attempts to demonstrate the presence of ferrous ions in blood plasma and extracellular fluids is another indication of their toxicity. As in cellular antioxidant systems, α-tocopherol is the primary lipid antioxidant in blood plasma.

Nutrients can influence oxygen radical damage as prooxidants, i.e., by contributing to the production of radicals of O_2 and PUFA, and as antioxidants, i.e., by suppressing oxygen radical generation, scavenging oxygen radicals, and catalyzing the degradation of their toxic reaction products. The following sections deal with this dual role of nutrients and with the potential for nutritional modulation of the pathology such radicals can cause in humans and animals. But it is first necessary to deal with the sources of oxygen radicals.

4.   Radicals of Molecular Oxygen

It may seem paradoxical that trauma, disease and death should be linked to the toxicity of O_2, which historically has been associated with the essence of life in higher organisms. Yet O_2 is lethal to anaerobic organisms, and it is now apparent that, in certain of its free radical forms, it is a cause of pathology in aerobes.

4.1   Superoxide Radicals

Nearly all the O_2 utilized by cells ($\geq 98\%$) is reduced to H_2O in the mitochondria by the addition of four electrons with the formation of ATP. However, this stepwise reduction of O_2 results in the release of small amounts of intermediate oxygen radicals. One such radical is the univalent reduction product superoxide (O_2^-). Although O_2^- is a relatively weak oxidant of uncertain toxicity, at high steady state levels it results in cessation of bacterial cell growth, mutagenesis and death (Fridovich, 1988). Organisms that lack SOD exhibit increased susceptibility to oxidants. Exposure of rats to hyperoxia increases SOD, catalase and GPx activity in the lungs, and induction of these enzymes by bacterial endotoxin administration enhances the resistance of this species to oxygen toxicity (Jamieson, 1988). SOD activity is highly variable among organs and species, and it is noteworthy that no increase in activity is seen in mice and guinea pigs, which do not develop resistance to hyperoxia. Hence SOD appears to have an important function in the prevention of oxygen toxicity.

The two most important reactions of O_2^-, nevertheless, appear to be its spontaneous dismutation to form H_2O_2:

$$O_2^- + O_2^- + 2H^+ \longrightarrow H_2O_2 + O_2$$

and its reduction of complexed ferric to ferrous ions that react with H_2O_2 to form hydroxyl radicals:

$$O_2^- + Fe^{3+} \longrightarrow Fe^{2+} + O_2$$

$$H_2O_2 + Fe^{2+} \longrightarrow HO\cdot + OH^- + Fe^{3+} \text{ (Fenton reaction)}$$

Sum: $O_2^- + H_2O_2 \longrightarrow O_2 + HO\cdot + OH^-$

(Iron-catalyzed Haber-Weiss reaction)

4.2   Hydroxyl Radicals

O_2 cannot be reduced by three electrons to $HO\cdot$ directly, but the iron-catalyzed reaction of O_2^- with H_2O_2 can, under conditions of high H_2O_2 production, be a major source of $HO\cdot$ radicals *in vivo*. Catalase, Se-GPx and SOD protect against some of the effects of toxicity. Besides being generated by dismutation of O_2^-, hydrogen peroxide is produced in several oxidase reactions by a two-electron reduction of diatomic oxygen. Although it has only a weak oxidant action toward most biomolecules, H_2O_2 is of major significance in oxygen radical toxicity because of its capacity to penetrate cell membranes rapidly and react with intracellular Fe^{2+}, Cu^+ and other transition metal ions to generate $HO\cdot$ radicals (Halliwell and Gutteridge, 1985). This is the apparent basis of its bactericidal action at high concentrations and may be one mechanism by which neutrophils kill invading organisms.

The cytochrome P-450 hydroxylase cycle in the endoplasmic reticulum is also a source of O_2^- and possibly $HO\cdot$ radicals. This system catalyzes the hydroxylation of xenobiotics, drugs, steroids and various other compounds at the expense of molecular oxygen:

$$RH + O_2 + 2 \text{ Cyt P-450-Fe}^{2+} + 2H^+ \longrightarrow ROH + H_2O + 2 \text{ Cyt-P-450-Fe}^{3+}$$

Binding of oxygen to cytochrome P-450-Fe^{2+} generates the intermediate RH-Fe^{3+}-O_2^- radical (Dolphin, 1988). Some dissociation of O_2^- from this complex occurs with consequent oxidative damage to lipids and other substances. This damage may be caused either by the protonated form of O_2^- ($HO_2 \cdot$) (Fridovich, 1988) or by dismutation of O_2^- to form H_2O_2 and subsequently HO· radicals.

Formation of HO· radicals from water is mainly responsible for the damage to cell membranes and DNA caused by ionizing radiation and is the putative cause of the associated development of cancer. Hydroxyl radicals react directly with DNA to form 8-hydroxyguanine and 8-hydroxy-2'-deoxyguanosine (Bertgold et al., 1988) and the excretion of these compounds in urine may provide an index of damage to nucleic acids caused by HO· radicals in vivo. Damage to lung tissue produced by exposure to ozone, on the other hand, is attributable largely to HO· radical-induced formation of lipoxy radicals, as indicated by the protective effect of lipid antioxidants. These two types of damage serve as examples of pathology caused directly by radicals of molecular oxygen and indirectly by their initiation of lipid peroxidation.

A positive side to the generally negative view of the role of oxygen radicals in biological systems is their participation in the microbial attack of neutrophils and phagocytes on invading microorganisms. Hydroxy radicals, along with various other oxidants including H_2O_2 and oxidized halogens, are produced, probably subsequent to O_2^- generation, by the "respiratory burst oxidase" present in these cells. Even here, however, leakage of oxy radicals from phagocytes causes damage to surrounding tissues (Halliwell and Gutteridge, 1985).

5.   Lipoxy Radicals

5.1   PUFA as Precursors

The PUFA located in cell membranes constitute the bulk of the lipids that are subject to peroxidation in the body. Their susceptibility to peroxidation differs markedly with the number of double bonds they contain and the prevalence of oxygen radicals, metal catalysts and antioxidants in the milieu of the membranes. The chemistry of lipid peroxidation in foods, and in biological systems in vitro and in vivo, has been extensively reviewed elsewhere (Sevanian and Hochstein, 1985; Chan, 1987; Bindoli, 1988) and will be discussed only briefly here.

The C-H bond energy of saturated fatty acids is sufficiently high to render them stable to oxidation under most circumstances, but bond energy at the α-position allylic to the olefinic groups of unsaturated fatty acids decreases markedly with increases in the number of double bonds. Consequently, propagation rates for fatty acids ranging from one to six double bonds increase in the order 0.025, 1, 2, 4, 6, 8 (Witting, 1965). The oils of fish and marine mammals contain high concentrations of pentaenoic and hexaenoic fatty acids that markedly increase their susceptibility to peroxidation, as well as that of the PUFA deposited in the tissues of animals to which they are fed.

The peroxidation of PUFA classically is depicted as a series of three or four basic reactions, but the process assumes increasing complexity with increases in the degree of unsaturation and the severity of peroxidative conditions.

Initiation: $RH \longrightarrow R\cdot$

O_2 uptake: $R\cdot + O_2 \longrightarrow ROO\cdot$

Propagation: $ROO\cdot + RH \longrightarrow ROOH + R\cdot$

Termination: $ROO\cdot + AH \longrightarrow ROOH + A\cdot$

The propagation (autoxidation) reaction is apparently rate-limiting. Termination may occur as the result of interactions between radicals to form stable products or of hydrogen abstraction from vitamin E or another lipid antioxidant to form hydroperoxides. Peroxidation also may be propagated by alkoxy radicals generated in the metal-catalyzed decomposition of hydroperoxides (see Section 9.3).

5.2   Fate of Dietary Peroxides

Fatty acid hydroperoxides formed in foods and edible oils can undergo extensive modification during processing and cooking. Decomposition produces aldehydes, ketones, hydrocarbons, esters and lactones. Further oxidation yields epoxides, dihydroperoxides, endoperoxides and keto derivatives. Condensation results in the formation of dimers and polymers, some of which decompose to form volatile products during heating (Frankel, 1987).

Peroxidized oils are not notably toxic when fed to animals. Fatty acyl hydroperoxides are reduced to hydroxy acids during absorption (Bergan and Draper, 1970) and the products of their decomposition in foods are apparently metabolized without adverse effects. Two mutagenic decomposition products, malondialdehyde (MDA) and hydroxynonenal (HNE) have been identified. MDA reacts with the ε-amino groups of lysine residues of food proteins, from which it is released during proteolysis and absorbed as a lysine adduct (Piché et al., 1988). HNE reacts avidly with the SH groups of sulfhydryl proteins (Esterbauer et al., 1988). High molecular weight products are largely unabsorbed.

5.3   Initiation of Lipid Peroxidation *in Vivo*

The events responsible for the initiation of lipid peroxidation *in vivo* are of central importance to the development of strategies for its control. Spontaneous peroxidation of PUFA is thermodynamically unfavorable. The initiation mechanism remains controversial, but there is strong evidence for an involvement of HO· radicals and transition metals. Stimulation of lipo-peroxide formation by Fe^{2+} ions is attributable to their reduction of H_2O_2 to form HO· radicals that, in turn, generate lipid radicals by abstracting hydrogen from the allylic carbons of PUFA. The stimulation caused by O_2^- is attributable to its dismutation to form H_2O_2. There are, nevertheless, numerous uncertainties surrounding the initiation process, including the source of the metal ions that participate in the Fenton reaction (Aust, 1988), evidence for initiation by a Fe^{2+}/Fe^{3+} complex (Braughler *et al.*, 1986), a possible role of O_2^- in the reduction of complexed Fe^{3+} ions and the signifi-cance of singlet oxygen ($^1O_2$) as an initiator of lipid peroxidation. These ambiguities are discussed in detail elsewhere (Halliwell, 1988b).

5.4   Fate of Lipid Peroxides Formed *in Vivo*

The products of PUFA hydroperoxides formed *in vivo* are less well defined. Hydroperoxides are reduced by the glutathione peroxidases to their corresponding alcohols, which are metabolized to CO_2 and H_2O via the gen-eral fatty acid oxidation pathway. Hydroperoxides also undergo decompo-sition *in vivo*, as demonstrated by the presence of HNE and MDA in the tissues and of MDA derivatives in the urine (Draper and Hadley, 1988). Exhalation of pentane and ethane (Tappel, 1982) attests to the peroxidative decomposition of n-6 and n-3 fatty acid peroxides, respectively.

Peroxidized fatty acids are formed *in vivo* primarily from PUFA located in the 2-position of membrane PL, from which they are cleaved by phospholipase A_2. An "oxidized phospholipid specific phospholipase A_2" has been detected in the peritoneal fluid of rats with induced inflammation (Inoue *et al.*, 1988). Cleavage of an oxidized PUFA is followed by reesterifi-cation with a normal fatty acid through the action of a lysophosphatide acyl-CoA transferase. The excised hydroperoxide is reduced to the alcohol analog by the glutathione peroxidases. An association has been observed between the presence of oxidized fatty acids in PL, increased membrane viscosity, decreased vesicle stability, and enhanced phospholipase A_2 activity (O'Brien, 1987).

In attempting to characterize the pathology associated with lipid per-oxidation *in vivo*, it is difficult to differentiate between lesions caused by disruption of the structure or function of cell membranes and those caused by reactions of the products of peroxidation with intracellular constituents.

Some gross effects of membrane disruption are evident in such phenomena as the hemolysis and release of bound lysosomal enzymes seen in vitamin E deficiency. However, lipids have a major role in regulating the transport of compounds across the outer and inner membranes of cells (Yeagle, 1989), and perturbation of this function may produce effects that are difficult to distinguish from those caused directly by the products of lipid peroxidation.

Some lipid peroxides have a role in physiological processes, such as the metabolism of prostaglandins, whereas others have cytotoxic effects that lead to a variety of pathologies. Possible chemical mechanisms underlying their pathological effects have been reviewed elsewhere (O'Brien, 1987). An early event following lipid peroxidation is an increase in membrane rigidity, which may impede the rotation of proteins within the lipid matrix and hence the transport of compounds into and out of the cell. The increased osmotic fragility and decreased deformability of aged erythrocytes appear to be manifestations of progressive oxidative changes in the cell membrane. The changes in fluidity could be due to the formation of *trans* double bonds, increased membrane polarity, polymerization of PUFA, crosslinking of oxidized lipids and proteins, protein crosslinking, or other events.

Addition of lipid peroxides to mitochondria rapidly inactivates respiratory enzymes, inhibits oxidative phosphorylation and causes loss of lipid and protein to the medium (O'Brien, 1987). Microsomes are also highly sensitive to small concentrations of lipid peroxides in the medium. The ability of lipid antioxidants to protect against many of these effects indicates that they are caused by fatty acyl peroxy radicals. Cultured hepatocytes are killed by *tert*-butyl hydroperoxide via a mechanism that depends on the direct oxidation of cellular lipids (Masaki *et al.*, 1989). Hemolysis caused by peroxy radicals is associated with oxidation of red blood cell membrane lipids and proteins (Niki *et al.*, 1988). Peroxides are capable of oxidizing the thiol groups of proteins, glutathione and other sulfhydryl compounds. Alkoxy radicals and other products of peroxide decomposition may also participate in these reactions. Linoleic acid hydroperoxides react with proteins to form fluorescent pigments similar to those found in the tissues. Such pigments may arise *in vivo* as the result of progressive autoxidation of PUFA and oxidation of proteins with condensation of the products to form polymers of increasing molecular weight (O'Brien, 1987).

There is clear evidence of biochemical damage caused by decomposition products of lipid peroxides formed *in vivo*. The hydroxynonenals (4-hydroxynonenal formed from n-6 PUFA and 4-hydroxyhexenal from n-3 PUFA) are highly reactive with the thiol groups of cysteine residues of proteins, as well as with their amino groups and those of phosphatidylserine and phosphatidylethanolamine. These aldehydes inactivate SH-enzymes *in vitro* and are the toxic metabolites of certain alkaloids ingested by humans and animals. Peroxidation of the LDL particle results in the formation of hydroxyalkenals that migrate to apoprotein B, modifying its ε-amino lysine groups and thereby impairing its uptake by the LDL receptor. 4-Hydroxynonenal also is genotoxic and mutagenic in cell cultures. The cytotoxic effects of hydroxyalkenals have been reviewed elsewhere (Esterbauer *et al.*, 1988).

As there is for oxidized fatty acids in PL, a hydrolytic mechanism exists for the recognition and elimination of oxidized proteins (Davies *et al.*, 1987. Oxidative denaturation of proteins is associated with a direct, quantitative increase in proteolytic susceptibility. A soluble proteolytic system has been identified in *E. coli* that can recognize and selectively degrade oxidatively denatured proteins (Davies and Lin, 1988). This system may protect cells against damage arising from the accumulation of denatured proteins, and may be regarded as a further means of defense against oxidative stress.

Malondialdehyde is a genotoxic and mutagenic product of lipid peroxidation formed *in vivo* and in foods by the decomposition of peroxides of fatty acids with three or more double bonds. Urinalysis has revealed the presence of MDA adducts with lysine, serine, ethanolamine and guanine, reflecting reactions *in vivo* with proteins, PL and nucleic acids (Draper and Hadley, 1988) The lysine-MDA adduct, which constitutes the main form of MDA absorbed from the diet and excreted in the urine (and possibly the main form occurring in the tissues) is non-mutagenic in the *Salmonella typhimurium* (Ames) test. Formation of Schiff's base adducts with other macromolecules may be a cause of oxygen radical pathology. For example, modification of apoprotein B by a reaction of MDA with the ε-amino groups of its lysine residues has been implicated in Watanabe heritable lipidemia of rabbits (Haberland *et al.*. 1988).

5.5   Role of Dietary Lipids

A question of major nutritional importance is whether there is a relationship between the level of PUFA in the diet, the generation of oxygen radicals in the body and the prevalence of oxygen radical pathology. This question has been approached in experimental animals by determining the effect of PUFA intake on the excretion of MDA adducts in urine. As for pentane exhalation (Tappel, 1982), MDA excretion in rats is increased under conditions which enhance lipid peroxidation, including vitamin E depletion and administration of catalysts such as iron and CCl_4 (Dhanakoti and Draper, 1987). Urinary MDA is also increased under conditions that stimulate lipolysis, including fasting, exercise, and administration of ACTH or epinephrine. Endogenous MDA excretion was unaffected, either in the fed or fasting state, by the level of corn oil feeding, indicating that *in vivo* lipid peroxidation was not increased by enrichment of the tissues with linoleic acid. However, fasting MDA excretion was increased by feeding highly unsaturated fatty acids from fish oil, signifying that a high intake of penta-enoic and hexaenoic fatty acids increased lipid peroxidation *in vivo*, at least under conditions of lipolysis. Feeding a diet high in salmon oil, as opposed to corn oil, has been observed to increase the concentration of lipofuscin pigments in the rat heart ventricle (Nalbone *et al.*, 1989).

There is a marked increase in lipid peroxides, measured as MDA, in rat liver and muscle homogenates following exhaustive exercise (Quintanilha 1988b). The cause of the increase in MDA excretion seen under various conditions of lipolysis is unknown and probably is not uniform. With respect to fasting and severe exercise, it is of interest that a 24-72 hr fast has been found to decrease hepatic catalase activity in rats (Godin and Wohaieb, 1988).

The relationship of PUFA intake to oxygen radical pathology in humans, if any, is as obscure as it is in animals. However, it is of interest that in several epidemiological studies of the relationship between dietary lipids and atherogenesis, the expected negative association between PUFA intake and serum cholesterol level was accompanied by a positive association with cancer incidence (Schatzkin *et al.*, 1988). The relevance of this obser-vation to free radical induction of carcinogenesis is highly speculative, but it is noteworthy that the advocacy of a high PUFA intake for the prevention of cardiovascular disease recently has been limited to a maximum of 10% of dietary energy (U.S. National Research Council, 1989).

6.   Vitamin E

The effect of PUFA intake on the oxidative stability of tissue lipids is heavily dependent upon the concurrent intake of vitamin E and other lipid antioxidants. The requirement for vitamin E can be reliably estimated using prediction formulas based on the amount of lipid consumed and its fatty acid composition (Witting, 1974). Depending upon these factors, the vitamin E requirement of animals can vary by one to two orders of magnitude.

A natural association between the PUFA and vitamin E content of fats and oils provides protection from risk of vitamin E deficiency at high PUFA intakes. For example, the inheritance of linoleic acid, the concentration of which differs substantially among varieties of corn, is linked to the inheritance of vitamin E, so that the linoleic acid : vitamin E ratio remains relatively constant. Similarly, the α-tocopherol content of rat liver subcellular fractions is correlated with their PUFA content, resulting in a similar ratio of these constituents in each fraction (Buttriss and Diplock, 1988). The highest known natural concentration of vitamin E occurs in the lipids of rubber latex (6-8% by weight) (Dunphy *et al.*. 1965; Chow *et al.*, 1969). A high concentration of vitamin E is necessary to maintain the oxidative stability of this highly unsaturated isoprenoid polymer which, when used to manufacture industrial products such as automobile tires, undergoes rapid oxidative deterioration unless it is stabilized by addition of synthetic antioxidants.

Fish oils are an exception to the general rule that the PUFA : vitamin E ratio of natural fats and oils is conducive to their oxidative stability in the diet and in the body. The highly unsaturated nature of fish oils and their relatively low vitamin E content render them unstable in experimental diets for animals, and feeding such oils is a classical method of inducing vitamin E deficiency. Their stability *in situ* may be due to the lower temperature and oxygen tension that prevail in fish tissues. Consumption of fish in the fresh or frozen state by Arctic native populations does not appear to have compromised their vitamin E status or resulted in any specific pathology (including cancer) that has been associated with oxygen radicals. Pharmaceutical preparations of n-3 fatty acids from fish oils are fortified with vitamin E to compensate for the increased requirement their consumption implies, and are encapsulated to protect them from oxidative deterioration.

The major diseases of vitamin E deficiency, exemplified by sterility in female rats and muscle dystrophy in rabbits, can be prevented and cured by administration of certain synthetic lipid antioxidants, among which N,N^1-diphenyl-p-phenylene diamine is the most active compound so far evaluated (Draper, 1980). This property is not indiscriminate among lipid antioxidants, as indicated by the fact that the common food antioxidants, BHA and BHT, are biologically inactive. The demonstration that antioxidants structurally unrelated to vitamin E are capable of replacing the vitamin *in vivo*, plus the fact that the metabolites of α-tocopherol formed *in vivo* are analogous to those produced by its oxidation in peroxidizing oils and by mild chemical oxidizing agents, indicate that the metabolism of vitamin E proceeds by chemical rather than by enzymatic processes. This raises the possibility that tissue concentrations above the normal physiological range may result in more efficient scavenging of lipoxy radicals, and forms the theoretical basis for the widespread practice of consuming vitamin E supplements for the prevention of cancer. While there is some evidence to indicate that a high intake of vitamin E may, in fact, reduce lipid peroxidation *in vivo*, the clinical significance of this effect is unknown.

Feeding a diet containing a high level of mixed antioxidants (300 ppm DL-α-tocopheryl acetate + 0.5% BHT + 0.5% vitamin C) suppressed MDA excretion by 35% in fasted rats previously fed a high PUFA diet (10% corn oil plus 5% cod liver oil) relative to that of controls fed a diet containing a normal level of vitamin E (30 ppm) (Dhanakoti and Draper, 1987). Since neither BHT nor vitamin C is a biologically active lipid antioxidant, the effect of the antioxidant mixture is attributable to its vitamin E content.

Administration of 1000 IU of vitamin E as d-α-tocopheryl acetate to human adults for 10 days has been reported to produce a 35% decrease in pentane exhalation (Lemoyne *et al.*, 1987). The plasma α-tocopherol level rose during supplementation from 0.77 to 1.46 mg/dL.

These studies indicate that lipid peroxidation *in vivo* is reduced in the presence of a high concentration of vitamin E in the tissues. However, the presence of MDA adducts with ethanolamine, serine and guanine in the urine of rats and humans (Draper and Hadley, 1988) shows that these compounds are formed in the course of normal physiological processes. Nevertheless, it would be of interest to determine the effect of vitamin E supplementation on the modification of these phospholipid and nucleic acid bases by MDA.

Chronic vitamin E deficiency has not been shown to increase cancer incidence in animals, and vitamin E is not an effective prophylactic for chemical carcinogenesis in animals. The protective effect of some synthetic antioxidants, including BHA and BHT, against chemically induced tumors appears to be due to stimulation of the mixed function oxidase system, which results in an increased rate of catabolism of concurrently administered carcinogens, rather than to the antioxidant properties of these compounds *per se* (Wattenberg, 1982). There is no cohort of human subjects with endemic vitamin E deficiency who could serve as a reference population as there is for vitamin A deficiency. Epidemiological evidence for a relationship between plasma vitamin E level and cancer prevalence within populations is inconsistent. The possible prophylactic effect of a vitamin E supplement is being evaluated in ongoing intervention trials (Simmons, 1986).

7.   Selenium

Selenium is ingested mainly in the form of selenomethionine and selenocysteine present in cereals. It occurs in Se-GPx as selenocysteine, which constitutes the active site on each of its four protein subunits. Inorganic forms of selenium, especially selenite, are capable of satisfying the dietary requirement for Se (though less efficiently than seleno amino acids), reflecting the ability of humans and animals to incorporate this element into selenocysteine.

Discovery of Se-GPx revealed an enzymatic catalysis of lipid peroxides and appeared to explain the nutritional interrelationship between Se and vitamin E. However, it was subsequently established that hydrogen peroxide, rather than lipid peroxides, is the substrate preferred by this enzyme, and that fatty acyl peroxides are reduced mainly by a non-Se-dependent glutathione peroxidase. In rat liver nuclei, for example, only 14% of total glutathione peroxidase activity is represented by the Se-enzyme (Tan *et al.*, 1988). A further non-Se-dependent GPx recently has been found in the soluble fraction of rat liver (Duan *et al.*, 1988). The protective action of Se-GPx against lipid peroxidation *in vivo*, therefore, is attributable largely to the catalysis of H_2O_2 reduction by glutathione (GSH), thereby preventing an oxidative attack on unsaturated lipids by HO· radicals generated in the Fenton reaction.

Functionally, Se-GPx has more in common with catalase than with the non-Se-GPx. There is marked variability in the relative concentrations of Se-GPx and catalase at different sites, but their combined activities in different tissues are relatively uniform (Halliwell and Gutteridge, 1985). Se-GPx activity can undergo a substantial decrease in Se deficiency before evidence of oxygen radical pathology is observed, indicating that there is normally excess capacity in the form of the two enzymes to prevent H_2O_2 accumulation. The high concentration of catalase in liver, erythrocytes and certain other tissues may be sufficient to prevent H_2O_2 accumulation. Nevertheless, from a comparative study of the enzymatic defense system against oxygen-derived free radicals in fibroblasts, it was concluded that the glutathione peroxidases have a key role (Michiels and Remacle, 1988).

Selenium deficiency is less inducive of lipid peroxidation *in vivo* than is vitamin E deficiency; indeed, it is difficult to demonstrate a deficiency of Se in experimental animals given excess vitamin E. However, muscle dystrophy in farm animals consuming rations limited in vitamin E content is epidemiologically related to the Se content of the soils on which their feeds are grown. Selenium deficiency produced experimentally in rats markedly increases their susceptibility to hyperoxia (Forman *et al.*, 1988). Paradoxically, the first unequivocal evidence for the essentiality of Se for the prevention of muscular dystrophy was provided by the finding that a fatal cardiomyopathy among children in the Keshan district of China (Keshan disease) is due to an extraordinarily low concentration of Se (about 0.005 ppm) in their indigenous diet (Yang *et al.*, 1984).

As for vitamin E, there is current interest in a possible relationship between Se intake and cancer incidence in humans. This interest arises from several sources: the role of Se-GPx in inhibiting oxygen radical generation; the finding that Se administration inhibits some forms of chemical carcinogenesis in experimental animals; epidemiological evidence for an inverse relationship between Se intake and cancer incidence in human populations. However, inhibition of chemical tumorigenesis requires near-toxic doses of Se, indicating that it may be due to a cytotoxic effect on cancer cells rather than to an increase in Se-GPx activity. The epidemiological evidence for cancer prophylaxis is inconsistent, and it is noteworthy that no unusual prevalence of cancer has been observed in those areas of China where Se deficiency has been endemic for many years. An intervention trial is in progress to test the effect on cancer incidence among U.S. adults of raising Se intake within the normal range (Clark and Combs, 1986). The magnitude of Se supplementation is constrained by a narrow margin of safety.

8.   Vitamin C

The role of vitamin C in the metabolism of oxygen radicals *in vivo* remains obscure. The best known biochemical action of ascorbate is its reduction of metal ions, exemplified in the increased absorption of food iron caused by its reduction to the ferrous state. Reduction of Fe^{2+} in the hydroxylase metalloenzymes involved in collagen synthesis also may explain the antiscorbutic activity of vitamin C, but this transformation can be effected *in vitro* by physiological reductants that lack antiscorbutic activity. It is not clear whether the synthesis of collagen is impaired in the scorbutic state to any greater extent than that of proteins which do not contain hydroxylated amino acids (England and Seifter, 1986).

Vitamin C is capable of scavenging O_2^- and HO· radicals and of reacting with singlet oxygen ($^1O_2$). Consequently, it has been postulated to inhibit oxygen radical damage *in vivo* by acting as a water-soluble antioxidant. Conceptually, this property might be important in tissues, such as the lens, that are low in SOD activity, or in lung tissue, which is exposed to gaseous oxidants (Halliwell and Gutteridge, 1985). There is evidence that vitamin C mitigates, to a limited extent, the effects of vitamin E deficiency in rats, but it is unlikely that this is due, as has been postulated, to reduction of the tocopheroxy radical formed in the antioxidant action of vitamin E, despite evidence that this reaction can occur in solution at high concentrations of ascorbate. However, vitamin C may exert an antioxidant effect by restoring oxidized thiols, such as glutathione, to the reduced state. Exposure of plasma to aqueous peroxy radicals has been observed to result in rapid oxidation of ascorbate, followed by the appearance of hydroperoxides of phospholipids, triglycerides and cholesterol esters (Frei *et al.*, 1988).

Paradoxically, the best known effect of ascorbic acid on oxygen radical metabolism is its stimulation of microsomal lipid peroxidation. This effect is due to generation of Fe^{2+} ions from complexes such as Fe^{3+}-ATP with consequent formation of HO· radicals in the Fenton reaction. The physiological significance of this *in vitro* action is, like many other *in vitro* observations on oxygen radical metabolism, questionable. Ascorbic acid has been reported to be toxic to iron-overload patients, but administration of 1 g per day to normal adults produces no increase in urinary MDA excretion, a sensitive indicator of increased lipid peroxidation *in vivo* (Draper, unpublished results).

9.   Iron

The role of iron in oxygen radical metabolism is, at the same time, obviously important and highly controversial. A brief account of this paradox follows.

9.1   Metabolism

About two-thirds of the iron in the body is present in hemoglobin and 10% in myoglobin. Smaller amounts are present in the transport protein transferrin, the soluble protein lactoferrin, various iron-containing enzymes, and in the iron storage proteins ferritin and hemosiderin, which are located mainly in the liver, spleen and bone marrow. Ferritin also occurs in plasma. Transferrin is transported across the outer cell membrane in a vacuole. Iron released into the cytoplasm forms loose chelates with several cell constituents including citrate, ADP and ATP. This soluble iron pool is drawn upon for the synthesis of iron proteins.

Overall iron balance is maintained by elaborate mechanisms of recovering iron released by hemoglobin catabolism, maintaining iron stores and enhancing the efficiency of iron absorption at chronically low intakes. Nevertheless, iron deficiency remains one of the most prevalent nutritional deficiency diseases, especially among populations consuming little animal protein. The deficiency is marked by depletion of iron stores, low plasma ferritin and transferrin saturation, and microcytic, hypochromic anemia.

Transferrin is normally only about 30% saturated with tightly bound iron, leaving a large reserve capacity for scavenging free iron. There is no mechanism for disposing of excess absorbed iron, and consequently the main defense against iron toxicity consists of a barrier to absorption located in the gut. Iron overload, which occurs in certain defects of iron absorption such as thalassemia and at very high dietary intakes, produces severe liver pathology. Partial protection from the toxic effects of iron administration is afforded by vitamin E, indicating that they are due in part to lipid peroxidation.

9.2   Role in Hydroxyl Radical Generation

Although it is tempting to attribute iron toxicity to catalysis of HO· generation in the Fenton reaction, the apparent absence of low molecular weight iron chelates in plasma and in other tissues ($\leq$ 5 μM) has raised a question regarding the form of iron involved in the generation of HO· radicals under normal physiological conditions. It is possible that, under steady state conditions, the rate of oxygen radical metabolism may be limited by the availability of iron in a form that is capable of undergoing redox reactions. Hydroxyl radical generation catalyzed by soluble iron complexes has been demonstrated in numerous *in vitro* experiments utilizing microsomes, but the concentration of iron generally used has been ten or more times its physiological levels (Halliwell, 1988b). The concentration and form(s) of iron that are capable of initiating a free radical chain reaction *in vivo* are still unknown.

Ferritin, transferrin and lactoferrin are not active catalysts of radical reactions leading to HO· formation. However, iron can be released from ferritin by O_2^- and from hemoglobin by H_2O_2 (Halliwell, 1988b). Trauma releases iron from soluble complexes in the intracellular pool, and it is therefore possible that, at least in some cases, the lipid peroxidation frequently associated with trauma is a result, rather than a cause, of tissue damage (Gutteridge, 1988). Large doses of vitamin E administered preoperatively have been reported to reduce the trauma associated with cardiopulmonary surgery in human subjects (Cararocchi *et al.*, 1988) and with brain and spinal cord injury in animals (Hall and Braughler, 1988). Enhanced lipid peroxidation following surgery may be due to release of iron from heme proteins by H_2O_2 or O_2^- with consequent HO· radical formation in the Fenton reaction, or to decomposition of peroxides by hemoglobin or myoglobin to form propagating alkoxy radicals.

It has been proposed that some form of iron produced in the Fenton reaction, rather than the HO· radical, is directly responsible for initiating lipid peroxidation (Aust, 1988). Inorganic iron exists in two common valencies, Fe^{2+} and Fe^{3+}, and in the form of the strongly oxidizing ferryl ion (Fe^{4+}) as well as the perferryl ion (Fe^{5+}). There is evidence that the initiation of peroxidation requires both Fe^{2+} and Fe^{3+} in a critical redox mixture or in some complex with an oxygen radical. There is also evidence that iron is capable of causing lipid peroxidation without the generation of HO· radicals. However, iron species have not been demonstrated to duplicate the reactions of HO· radicals with aromatic compounds, to respond similarly to HO· scavengers, or to yield the same spin adducts, nor has any active iron complex been characterized (Halliwell, 1988b).

9.3   Role in Lipid Peroxide Decomposition

In addition to its role in HO· generation, iron can enhance chain reactions among unsaturated lipids by catalyzing the decomposition of preformed fatty acyl hydroperoxides:

$$ROOH + Fe^{2+} \longrightarrow RO\cdot + OH^- + Fe^{3+}$$
$$ROOH + Fe^{3+} \longrightarrow ROO\cdot + H^+ + Fe^{2+}$$

The alkoxy and peroxy radicals produced in these reactions propagate further chain reactions. Iron catalysis of lipid peroxidation by this mechanism is readily demonstrable in unsaturated oils, but it is unclear whether sufficient concentrations of unsequestered Fe^{2+} or Fe^{3+} ions are available to catalyze similar reactions *in vivo*. Conversely, it has been suggested that, instead of acting as an initiator of lipid peroxidation *in vitro*, iron may stimulate peroxidation by decomposing preformed peroxides invariably present in liposomal and microsomal preparations (Gutteridge, 1988).

"Oxidized flavors" caused by aldehydes, ketones, furans, lactones and other products of iron-catalyzed decomposition of fatty acid peroxides are a major problem in the food industry. Catalysis by ionic forms of iron, copper, cobalt and other transition elements occurs during food processing. Catalysis of peroxidation by heme iron is particularly prevalent in poultry products prepared by mechanical processes, which frequently cause fractures of small bones. Animal products are more susceptible to peroxidation than are plant foods because they contain more highly unsaturated ($\geq$ tetraenoic) fatty acids and more active iron catalysts. The general order of catalytic activity is hematin > hemoproteins > Fe^{2+} > Fe^{3+}.

A protective side to the involvement of iron in oxygen radical metabolism is reflected in its presence in catalase and in the iron-containing superoxide dismutase present in some bacteria and plants. Iron deficiency, either in the mild form that is common in some human populations or in an acute form in experimental animals, has no apparent effect on the activity of catalase and other iron-containing enzymes or on oxygen radical metabolism. The rigor with which these enzymes are conserved attests to their physiological importance in the metabolism of molecular oxygen and its radicals.

10. Copper, Zinc and Manganese

The metal-containing dismutases represent part of the defense system against cell damage caused, directly or indirectly, by O_2^- in eukaryotes. One dismutase, containing both Cu and Zn, is found predominantly in the cytosol and to a limited extent in extracellular fluids. The other, containing Mn, is found in mitochondria. MnSODs and a family of FeSODs are characteristic of prokaryotes, whereas (except for a small subset of bacteria) CuZnSODs are limited to eukaryotes. All SODs catalyze the same dismutation reaction with comparable efficiency.

Copper undergoes alternating reduction and oxidation in the course of the SOD reaction; Zn does not function directly in catalysis but is essential for stabilization of the enzyme. Dietary Cu, Zn and Mn are obviously essential for maintaining this component of cellular defenses against damage caused by radicals of molecular oxygen and, consequently, of lipoxy radicals.

10.1   Copper

Liver microsomes and mitochondria from Cu-deficient rats exhibit increased NADPH-induced lipid peroxidation *in vitro*, but there is no clear evidence that Cu deficiency increases peroxidation *in vivo*. Despite a 90% reduction in CuZnSOD activity observed in the liver of Cu-deficient rats relative to that of pair-fed or ad libitum-fed controls, there were no differences in the primary free radical defense system that could be attributed to Cu deficiency specifically (Taylor *et al.*, 1988). Similar observations were made with respect to lung and heart (Bettger and Bray, 1989). CuZnSOD activity in rat liver is increased by Cu loading; it is unclear whether this increase is due to induction or to greater saturation of the enzyme.

The lack of evidence of increased lipid peroxidation in the tissues of rats subjected to acute Cu deficiency, despite a marked reduction in CuZnSOD, implies that this enzyme is not of major importance in preventing the generation of HO· radicals, presumed initiators of peroxidation. It is noteworthy that depletion of CuZnSOD is associated with increases in catalase and the two peroxidases that decompose H_2O_2, thereby preventing HO· radical generation in the iron-catalyzed Haber-Weiss reaction. The importance of these enzymes, relative to that of SOD, in the prevention of O_2^--driven HO· formation is further indicated by the fact that O_2^- spontaneously dismutates to form H_2O_2 under physiological conditions.

In contrast to the failure of CuZnSOD depletion caused by Cu deficiency to affect the oxidative stability of lipids, depletion of Se-GPx by Se deficiency produces clear evidence of increased *in vivo* lipid peroxidation in experimental animals, particularly when the intake of vitamin E is also restricted. These observations underscore the importance of the H_2O_2 metabolizing enzymes in preventing lipid peroxidation. They militate against a major role of SOD enzymes as inhibitors of lipid peroxidation and other forms of oxygen radical damage. The pathology of Cu, Zn and Mn deficiency diseases is not indicative of tissue damage caused by HO· radicals. This does not preclude the possibility that there is other, unrecognized pathology caused by superoxide.

Most of the reactions involving oxygen radicals that are catalyzed by iron, including the Fenton reaction and the decomposition of fatty acyl hydroperoxides, also can be catalyzed by copper ions. Cu^{2+} salts produce HO· radicals in the presence of H_2O_2 and a suitable reducing agent. Superoxide is an effective Cu^{2+} reductant. Physiological ligands of Cu^{2+} ions, such as albumin and histidine, do not fully prevent this reduction. Copper contamination of industrial oils and processed foods is a common cause of oxidative rancidity. Most plasma copper is present in ceruloplasmin, but this protein does not sequester and release copper in the manner that transferrin binds and releases iron. In Wilson's disease, an inherited defect marked by low plasma ceruloplasmin, deposition of copper in the tissues is associated with evidence of free radical pathology. Under normal physiological conditions, Cu^+ ions appear to be bound mainly by histidine and albumin in plasma, and by GSH (Freedman et al., 1989), metallothioneins and other proteins intracellularly.

10.2   Zinc

Zinc deficiency causes increased free radical generation in rat lung microsomes, as measured by electron spin-trapping techniques (Kubow et al., 1986), as well as increased liver microsomal NADPH-induced lipid peroxidation and H_2O_2 production (Taylor et al., 1988). However, even acute Zn deficiency has little effect on the tissue concentrations of this element and, presumably for this reason, CuZnSOD activity remains unimpaired. The lungs of Zn deficient rats exhibit increased CuZnSOD activity and an increased concentration of copper. Although a decrease in liver metallothionein has been seen in Zn deficient animals, no impairment of the primary antioxidant defense system was observed that was not attributable to inanition (Taylor et al., 1988).

10.3   Manganese

Mitochondrial MnSOD activity in liver is decreased in Mn deficiency but, as in Zn deficiency, there is a compensating increase in CuZnSOD activity. MnSOD activity varies markedly among organs and species, as does the effect of Mn deficiency on other components of the antioxidant defense system. The importance of MnSOD in the metabolism of oxygen radicals by eukaryotic cells, as well as the mechanism of its action (except that it involves a redox change in Mn) remain unclear. Increased MnSOD activity in the pulmonary macrophages of newborn rats exposed to hyperoxia, and decreased activity in a pathogen-free environment, suggest an important function in neonates. Newborn rats adapt more rapidly than adults to hyperoxia.

11.   Sulfur Amino Acids

Sulfur amino acids (SAA) have a mild "sparing effect" on the vitamin E requirement of animals. For example, the experimental production of nutritional muscular dystrophy in the chick requires a diet that is limiting in SAA as well as in vitamin E and Se; otherwise the animal develops encephalomalacia (vitamin E deficiency) or exudative diathesis (vitamin E and Se deficiency). This finding presumably is attributable to the need for the synthesis of sufficient reduced glutathione (GSH) to fuel the reduction of lipid hydroperoxides and H_2O_2 by the glutathione peroxidases:

$$ROOH + 2GSH \longrightarrow ROH + GSSG + H_2O$$

$$H_2O_2 + 2GSH \longrightarrow GSSG + H_2O$$

Most of the free glutathione in the tissues is present in the reduced rather than the oxidized state (GSSG), but a substantial portion occurs in disulfide linkages with other SH compounds. Exposure of human blood plasma to aqueous peroxyl radicals results in depletion of SH groups as an early event (Frei et al., 1988). The GSH redox system appears to be a necessary adjunct to vitamin E in the protection of hepatocytes from oxidative stress (Pascoe and Reed, 1989). SAA deficiency increases the susceptibility of rats to hyperoxia (Forman et al., 1988).

The tissue level of GSH is maintained by glutathione reductases, which catalyze the reduction of GSSG and other disulfides by NADPH generated in the pentose shunt:

$$GSSG + NADPH^+ + H^+ \longrightarrow 2GSH + NADP^+$$

12. Niacin

NADPH, a coenzyme form of niacin, is therefore another nutrient-dependent factor required for the prevention of oxygen radical pathology. Even when adequate niacin is available from the diet or from biosynthesis, an impaired supply of glucose to a tissue may restrict the availability of NADPH reducing equivalents for the synthesis of reduced glutathione. This may be the case, for example, in the oxidative damage to the lens that occurs in glaucoma (Halliwell and Gutteridge, 1985).

13. β-Carotene

In addition to its role as a provitamin A, β-carotene is an avid quencher of singlet oxygen ($^1O_2$) which, although it is not an oxygen radical, is capable of carrying out many of the same reactions. β-carotene can be an important stabilizer of oxidizable compounds, including PUFA, in foods of plant origin, particularly under conditions of exposure to light or ionizing radiation that generate singlet oxygen. Judging from their high content, it may also function in this role in the eyes of some species. β-carotene is capable of scavenging free radicals, and has been reported to be effective in the treatment of erythropoietic protoporphyria and other diseases of photosensitivity (Mathews-Roth, 1985). It can be a significant source of lipid antioxidant activity in blood plasma (Burton and Ingold, 1984) and possibly in some subcellular membranes.

There is persuasive epidemiological evidence for a negative association between β-carotene intake or plasma level and cancer incidence at several sites. However, it is possible that β-carotene is simply a marker for a diet subculture, high in fruits and vegetable and low in fat, that is protective against cancer for other reasons. β-carotene is not an essential nutrient. Circumpolar populations have maintained nutritional health and shown no unusual predisposition to cancer over many generations despite negligible amounts of β-carotene in their diet. It is unlikely, therefore, that within the normal range of intakes, β-carotene has a significant effect on cancer risk. Intervention trials on the value of pharmacological intakes of β-carotene in cancer prophylaxis are in progress (Simmons, 1986).

14. Overview

The risk of human pathology arising from a dietary deficiency of the nutrients involved in the metabolism of oxygen radicals appears, on the basis of current evidence, to be small. Copper and Mn deficiency are not significant issues in human nutrition and, judging from the results of studies on animals, deficiencies of these elements, in any event, would not seriously impair the antioxidant defense system. While there is some evidence of mild Zn deficiency in some cohorts of the human population, animal experimentation indicates that no significant impairment of oxygen radical metabolism occurs in the deficient state other than that caused by inanition. Abundant evidence is available for pathology arising from iron catalysis of oxygen radical generation in the tissues of humans and animals subjected to various forms of trauma, but except for cases of iron overload from non-food sources and of impaired control of iron absorption, there is little evidence of pathology associated with the normal range of intakes of this element in the human diet.

The high protein diet consumed in industrialized countries contains a surfeit of the sulfur amino acids required for synthesis of the thiol compounds involved in the metabolism of oxy radicals, including the glutathione required to sustain peroxidase activity. This diet also provides adequate amounts of riboflavin and niacin equivalents to maintain glutathione reductase activity. When the energy requirement is satisfied, the lower protein diets of cereal-based food economies also are capable of meeting the needs for these nutrients. However, inanition and starvation result in depletion of antioxidant enzymes and increased production of oxygen radicals (Godin and Wohaieb, 1988).

Selenium deficiency is limited mainly to farm animals fed feedstuffs grown on Se deficient local soils. The cardiomyopathy of Keshan disease in humans, however, appears to be precipitated by consumption of foodstuffs grown on local Se deficient soils. The Se status of some national populations is substantially lower than that of others because of differences in the level of Se in the general food supply. Residents of New Zealand have significantly lower plasma Se and Se-GPx levels than those of most other countries, yet there is no epidemiological evidence that these levels are associated with oxygen radical pathology or an increased incidence of cancer or other diseases.

Apart from cases of malabsorption or defective transport, vitamin E deficiency is not a recognized clinical entity in human nutrition. There are no major differences in vitamin E status, as there are for selenium, among populations. Consequently, interpopulation epidemiology is not useful as a source of information relative to any possible relationship between dietary vitamin E and oxygen radical pathology.

Vitamin E supplements, on the other hand, have been demonstrated to inhibit the tissue damage that occurs in various conditions associated with the generation of oxygen radicals, including exposure to hyperoxia, xenobiotics and gaseous oxidants, iron overload, postsurgical trauma, brain or spinal cord injury, reperfusion injury and inflammation. There is also evidence that large doses of vitamin E reduce lipid peroxidation in normal humans and animals. These findings justify further research into the effect of pharmacological levels of vitamin E on the prevention and clinical management of pathologies associated with increased oxygen radical production, as well as the effect of dietary vitamin E intake on the prevention of oxidative damage to DNA and other sensitive macromolecules.

Despite a prevalent public concern about vitamin C adequacy, intakes of this vitamin from normal diets exceed the amount required to prevent all biochemical and clinical signs of deficiency. In general, human plasma levels of vitamin C exceed those of animals that synthesize the vitamin (Sheahan, 1947) and claims for a prophylactic or therapeutic effect of megadoses of the vitamin on various disease states, including cancer, are unsubstantiated. As a reducing agent, vitamin C has been postulated to play various physiological roles in oxygen radical metabolism, but most of the evidence for these hypotheses is based on *in vitro* observations of dubious relevance to *in vivo* events. Unlike those of vitamin E deficiency, the symptoms of vitamin C deficiency are not indicative of an aberration in oxygen radical metabolism. It is highly unlikely, therefore, that dietary vitamin C deficiency is a factor in the occurrence of diseases mediated by these radicals.

In assessing the role of nutrients in the prevention or modulation of diseases associated with oxygen radical pathology, it is important to distinguish between their use as supplements to the diet, designed to reduce the risk of a nutritional deficiency, and their use as prophylactic drugs. The prevalent practice of taking megadoses of β-carotene, vitamin C or vitamin E is clearly a pharmacological use of these nutrients, fuelled mainly by their postulated inhibition of oxygen radical-induced carcinogenesis. Their use as drugs is made possible by their relative lack of toxicity, attributable mainly to poor absorption of β-carotene, to a low renal threshold for vitamin C, and to a limited binding of vitamin E to lipoproteins that prevents its accumulation in the blood to about twice the normal level. Comparable intakes of Se and vitamin A (two other nutrients implicated in cancer prevention) are precluded by their narrow margins of safety.

Whether megadoses of β-carotene, vitamin C or vitamin E reduce the risk of cancer and other pathologies in which oxygen radicals have been implicated is a pharmacological question that, for purposes of intervention strategy, should be distinguished from the nutritional question whether the risk of such diseases is related to the habitual intake of these vitamins in the diet (Draper, 1988). Nutrients used as drugs should be subject to the same standards of efficacy and safety as other drugs. Food fortification with nutrients as a means of prophylaxis for chronic degenerative diseases would distort the meaning of nutrition recommendations, undermine public confidence in the adequacy of the food supply, and preclude selectivity in target populations. Should prophylaxis with a specific nutrient be found efficacious and safe in the case of a specific disease, the use of supplements to the diet would be preferable to food fortification from the standpoint of compliance, selectivity and the integrity of diet recommendations for the prevention of nutrient deficiencies.

References

Aust, S. D., 1988, Sources of iron for lipid peroxidation in biological systems, in: *Oxygen Radicals and Tissue Injury* (B. Halliwell, ed.), pp. 27-33. Federation of American Societies for Experimental Biology, Bethesda, MD.

Bergan, J. G., and Draper, H. H., 1970, Absorption and metabolism of 1-^{14}C-methyl linoleate hydroperoxide, *Lipids* 5:976.

Bertgold, D. S., Simic, M. G., Alessio, H., and Cutler, R. G., 1988, Urine markers for oxidative DNA damage, in: *Oxygen Radicals in Biology and Medicine* (M. G. Simic, K. A. Taylor, J. E. Ward, and C. von Sonntag, eds.) pp. 483-489, Plenum Press, New York.

Bettger, W. G., and Bray, T. M., 1989, Effect of dietary zinc or copper deficiency on catalase, glutathione peroxidase and superoxide dismutase activities in rat heart, *Nutr. Res.* **9**:319.

Bindoli, A., 1988, Lipid peroxidation in mitochondria, *Free Rad. Biol. Med.* **5**:247.

Braughler, J. M., Duncan, L. A., and Chase, R. L., 1986, The involvement of iron in lipid peroxidation; importance of ferric to ferrous ratios in initiation, *J. Biol. Chem.* **261**:1082.

Burton, G. W., and Ingold, K. V., 1984, Beta carotene: an unusual type of lipid antioxidant, *Science* **224**:569.

Buttriss, J. L., and Diplock, A. T., 1988, The α-tocopherol and phospholipid fatty acid content of rat liver subcellular membranes in vitamin E and selenium deficiency, *Biochim. Biophys. Acta* **963**:61.

Cararocchi, N. C., England, M. D., O'Brien, J. F., Solis, E., Russo, P., Schaff, H. V., Orszulak, T. A., Pluth, J. R., and Kaye, M. P., 1988, Superoxide generation during cardiopulmonary bypass: is there a role for vitamin E? *J. Surg. Res.* **30**:519.

Chan, H. W.-S, (ed.), 1987, *Autoxidation of Unsaturated Lipids*, Academic Press, New York.

Chow, C. K., (ed.), 1988, *Cellular Antioxidant Defense Mechanisms*, vol. 1, 2 and 3, CRC Press, Boca Raton, Fl.

Chow, C. K., Draper, H. H., and Csallany, A. S., 1969, Method for the assay of free and esterified tocopherols, *Anal. Biochem.* **32**:81.

Clark, L. C., and Combs, G. F., Jr., 1986, Selenium compounds and the prevention of cancer: research needs and public health implications, *J. Nutr.* **116**:170.

Davies, K. J. A., and Lin, S. W., 1988, Degradation of oxidatively denatured proteins in *Escherrichia coli*, *Free Rad. Biol. Med.* **5**:215.

Davies, K. J. A., Lin, S. W., and Pacifici, R. E., 1987, Protein damage and degradation by oxygen radicals, IV. Degradation of denatured protein, *J. Biol. Chem.* **262**:9914.

Dhanakoti, S. N., and Draper, H. H., 1987, Response of urinary malondialdehyde to factors that stimulate lipid peroxidation *in vivo*, *Lipids* **22**:643.

Dolphin, D., 1988, The generation of radicals during the normal and abnormal functioning of cytochromes P-450, in: *Oxygen Radicals in Biology and Medicine* (M. G. Simic, K. A. Taylor, J. F. Ward, and C. von Sonntag, eds.), pp. 491-500, Plenum Press, New York.

Draper, H. H., 1980, Role of vitamin E in plants, microbes, invertebrates and fish, in: *Vitamin E, a Comprehensive Treatise* (L. J. Machlin, ed.), pp. 391-396, Marcel Dekker, New York.

Draper, H. H., 1988, Nutrients as nutrients and nutrients as prophylactic drugs, *J. Nutr.* **118**:1420.

Draper, H. H., and Bird, R. P., 1984, Antioxidants and cancer, *J. Agr. Food Chem.* **32**:433.

Draper, H. H., and Hadley, M., 1988, Malondialdehyde derivatives in urine, in: *Oxygen Radicals in Biology and Medicine* (M. Simic, K. A. Taylor, J. F. Ward, and C. von Sonntag, eds.), pp. 199-202, Plenum Press, New York.

Duan, Y-J., Komura, S., Fiszer-Szafarz, B., Szafarz, D., and Yagi, K., 1988, Purification and characterization of a novel monomeric glutathione peroxidase from rat liver, *J. Biol. Chem.* **263**:19003.

Dunphy, P. J., Whittle, K. J., Pennock, J. F., and Morton, R. A., 1965. Identification and estimation of tocotrienols in *Hevea* latex, *Nature* **207**:521.

Englard, S., and Seifter, S., 1986, The biochemical functions of ascorbic acid, *Ann. Rev. Nutr.* **6**:365.

Esterbauer, H., Zollner, H., and Schaur, R. J., 1988, Hydroxyalkenals: cytotoxic products of lipid peroxidation, *ISI Atlas of Science: Biochemistry* **1**:311.

Forman, H. J., Skelton, D. C., Loeb, G. A., and Dorio, R. J., 1988, Membrane permeability and oxidant induced injury, in: *Oxygen Radicals in Biology and Medicine* (M. Simic, K. A. Taylor, J. F. Ward, and C. von Sonntag, eds.), pp. 523-530, Plenum Press, New York.

Frankel, E. N., 1987, Secondary products of lipid oxidation, *Chem. Phys. Lipids* **44**:73.

Freedman, J. H., Ciriolo, M. R., and Peisach, J., 1989, The role of glutathione in copper metabolism and toxicity, *J. Biol. Chem.* **264**:5598.

Frei, B., Stocker, R., and Ames, B. N., Antioxidant defenses and lipid peroxidation in human blood plasma, *Proc. Nat. Acad. Sci., USA* **85**:9748.

Fridovich, I., 1988, The biology of oxygen radicals: general concepts, in: *Oxygen Radicals and Tissue Injury* (B. Halliwell, ed.), pp. 1-5, Federation of American Societies for Experimental Biology, Bethesda, MD

Godin, D. V., and Wohaieb, S. A., 1988, Nutritional deficiency, starvation, and tissue antioxidant status, *Free Rad. Biol Med.* **5**:165.

Gutteridge, J. M. C., 1988, Lipid peroxidation: some problems and concepts, in: *Oxygen Radicals and Tissue Injury* (B. Halliwell, ed.), pp. 9-19, Federation of American Societies for Experimental Biology, Bethesda, MD.

Haberland, M. E., Fong, D., and Cheng, L., 1988, Malondialdehyde-altered protein occurs in atheroma of Watanabe heritable hyperlipidemic rabbits, *Science* **241**:215.

Hall, E. D., and Braughler, J. M., 1988, The role of oxygen radical-induced peroxidation in acute central nervous system trauma, in: *Oxygen Radicals and Tissue Injury* (B. Halliwell, ed.), pp. 92-98, Federation of American Societies for Experimental Biology, Bethesda, MD.

Halliwell, B., (ed.), 1988a, *Oxygen Radicals and Tissue Injury, Proc. Brook Lodge Symposium, Augusta, Michigan, April 27-29, 1987*, 148 pp., Federation of American Societies for Experimental Biology, Bethesda, MD.

Halliwell, B., 1988b, A radical approach to human disease, in: *Oxygen Radicals and Tissue Injury* (B. Halliwell, ed.), pp. 139-143, Federation of American Societies for Experimental Biology, Bethesda, MD.

Halliwell, B., and Gutteridge, J. M. C., 1985, *Free Radicals in Biology and Medicine*, 346 pp., Clarendon Press, Oxford.

Inoue, K., Itabe, H., and Kudo, I., 1988, Production and metabolism of cytotoxic phospholipids generated during incubation of liposomes with oxyhemoglobin, in: *Oxygen Radicals in Biology and Medicine* (M. Simic, K. A. Taylor, J. F. Ward, and C. von Sonntag, eds.), pp. 291-299, Plenum Press, New York.

Jamieson, D., 1988, Reactive oxygen metabolites and hyperbaric toxicity, in: *Oxygen Radicals in Biology and Medicine* (M. Simic, K. A. Taylor, J. F. Ward, and C. von Sonntag, eds.), pp. 553-560, Plenum Press, New York.

Kubow, S., Bray, T. M., and Bettger, W. J., 1986, Effects of dietary zinc and copper on free radical production in rat lung and liver, *Can. J. Physiol. Pharmacol.* **64**:1281.

Lemoyne, M., Van Gossum, A., Kurian, R., Ostro, M., Axler, J., and Jeejeebhoy, K. N., 1987, Breath pentane analysis as an index of lipid peroxidation: a functional test of vitamin E status, *Am. J. Clin. Nutr.* **46**:267.

Masaki, N., Kyle, M. E., and Farber, J. L., 1989, *tert*-Butyl hydroperoxide kills cultured hepatocytes by peroxidizing membrane lipids, *Arch. Biochem. Biophys.* **269**:390.

Mathews-Roth, M. M., 1986, Beta carotene therapy for erythropoietic protoporphyria and other photosensitivity diseases, *Biochimie* **68**:875.

Michiels, C., and Remacle, J., 1988, Microinjection of antioxidant enzymes to protect cells from oxygen derived free radicals, in: *Oxygen Radicals in Biology and Medicine* (M. Simic, K. A. Taylor, J. F. Ward, and C. von Sonntag, eds.), pp. 703-711, Plenum Press, New York.

Nalbone, G., Leonardi, J., Termine, E., Portugal, H., Lechene, P., Pauli, A.-M., and Lafont, H., Effects of fish oil, corn oil and lard diets on lipid peroxidation status and glutathione peroxidase activities in rat heart, *Lipids* **24**:179.

Niki, W., Yamamoto, Y., Takahashi, M., Yamamoto, K., Yamamoto, Y., Komuro, E., Miki, M., Yasuda, H., and Mino, M., 1988, Free radical-induced damage of blood and its inhibition by antioxidants, *J. Nutr. Sci. Vitaminol.* **34**:507.

O'Brien, P. J., 1987, Oxidation of lipids in biological membranes and intracellular consequences, in: *Autoxidation of Unsaturated Lipids* (H. W-S. Chan, ed.), pp. 233-280, Academic Press, New York.

Pascoe, G. A., and Reed, D. J., 1989, Cell calcium, vitamin E, and the thiol redox system in cytotoxicity, *Free Rad. Biol. Med.* **6**:209.

Piché, L. A., Cole, P. D., Hadley, M., van den Bergh, R., and Draper, H. H., 1988, Identification of N-e-(2-propenol) lysine as the main form of malondialdehyde in food digesta, *Carcinogenesis* **9**:473.

Quintanilha, A., (ed.), 1988a. *Reactive Oxygen Species in Chemistry, Biology, and Medicine*, 232 pp., Plenum Press, New York

Quintanilha, A. T., 1988b, Oxidative effects of physical exercise, in: *Reactive Oxygen Species in Chemistry, Biology and Medicine* (A. Quintanilha, ed.), pp. 187-195, Plenum Press, New York.

Schatzkin, A., Hoover, R. N., Taylor, P. R., Ziegler, R. G., Carter, C. L., Albanes, D., Larson, D. B., and Licitra, L. M., 1988, Site-specific analysis of total serum cholesterol and incident cancer in the National Health and Nutrition Examination Survey, I. Epidemiologic follow-up study, *Cancer Res.* **48**:452.

Sevanian, A., and Hochstein, P., 1985, Mechanisms and consequences of lipid peroxidation in biological systems, *Ann. Rev. Nutr.* **5**:365.

Sheahan, M-M., 1947, The ascorbic acid content of the blood serum of farm animals, *J. Comp. Path.* **57**:28.

Simic, M. G., Taylor, K. A., Ward, J. F., and von Sonntag, C., eds., 1988, *Oxygen Radicals in Biology and Medicine, Proc. Fourth Intern. Congress on Oxygen Radicals,* June 27-July 3, 1987, LaJolla, California, 1085 pp., Plenum Press, New York.

Simmons, K., 1986, Evaluating vitamin prophylaxis for cancer, *JAMA* **265**:1832.

Tan, K. H., Meyer, D. J., Gillies, N., and Ketterer, B., 1988, Detoxification of DNA hydroperoxide by glutathione transferases and the purification and characterization of glutathione transferases of the rat liver nucleus, *Biochem. J.* **254**:841.

Tappel, A. L., 1982, Measurement of *in vivo* lipid peroxidation via exhaled pentane and protection by vitamin E, in: *Lipid Peroxides in Biology and Medicine* (K. Yagi, ed.). pp. 213-222, Academic Press, New York.

Taylor, C. G., Bettger, W. J., and Bray, T. M., 1988, Effect of dietary zinc or copper deficiency on the primary free radical defense system in rats, *J. Nutr.* **118**:613.

U. S. National Research Council, Committee on Diet and Health, Food and Nutrition
    Board, Commission on Life Sciences, 1989, Diet and Health, National Academy
    Press, Washington, DC, 20 pp.
Wattenberg, L. W., 1982, Inhibition of chemical carcinogens by minor dietary
    components, in: *Molecular Interrelations of Nutrition and Cancer* (M. S. Arnott,
    J. van Eys, and Y. M. Wang, eds.), Raven Press, New York, pp. 43-56.
Witting, L. A., 1965, Lipid peroxidation *in vivo*, *J. Am. Oil Chem. Soc.* **42**:908.
Witting, L. A., 1974, Vitamin E-polyunsaturated lipid relationship in diet and tissues,
    Am. J. Clin. Nutr. **27**:952.
Yang, G., Chang, J., Wen, Z., Ge, K., Zhu, L., Chen, X., and Chen, X., 1984, The role
    of selenium in Keshan disease, *Adv. Nutr. Res.* **6**:203.
Yeagle, P. L., 1989, Lipid regulation of cell membrane structure and function, *FASEB
    J.* **3**:1833.

Index